P9-EEH-057

Facility Manager's Guide to Security:

Protecting Your Assets

11294

MAR 0 1 2008

TH
9705
.R45
2005

Facility Manager's Guide to Security:

Protecting Your Assets

By
Robert N. Reid

INFORMATION RESOURCES CENTER
ASIS INTERNATIONAL
1625 PRINCE STREET
ALEXANDRIA, VA 22314
TEL. (703) 519-6200

THE FAIRMONT PRESS, INC.

DEKKER
CRC PRESS

Library of Congress Cataloging-in-Publication Data

Reid, Robert N., 1949-
Facility manager's guide to security : protecting your assets / by Robert N. Reid.
    p. cm.
    Includes bibliographical references and index.
    ISBN: 0-88173-479-9 (print) — 0-88173-483-7 (electronic)
1. Buildings—Security Measures. I. Title

    TH9705.R45 2005
    658.4'73--dc22

                2004061943

Facility manager's guide to security : protecting your assets by Robert N. Reid. ©2005 by The Fairmont Press, Inc. All rights reserved. No part of this publication may be reproduced or transmitted in any form or by any means, electronic or mechanical, including photocopy, recording, or any information storage and retrieval system, without permission in writing from the publisher.

Published by the Fairmont Press, Inc.
700 Indian Trail
Lilburn, GA 30047
tel: 770-925-9388; fax: 770-381-9865
http://www.fairmontpress.com

Distributed by Marcel Dekker/CRC Press
2000 N.W. Corporate Blvd.
Boca Raton, FL 33431
tel: 800-272-7737
http://www.crcpress.com

Printed in the United States of America
10 9 8 7 6 5 4 3 2 1

0-88173-479-9 (The Fairmont Press, Inc.)
0-8247-2162-4 (Marcel Dekker/CRC Press)

While every effort is made to provide dependable information, the publisher, authors, and editors cannot be held responsible for any errors or omissions.

For Kathy, July 30, 2004

Table of Contents

Foreword

By
Joseph F. Gustin,
Author of *Disaster and Recover Planning*, Fairmont Press 2004.

Robert Reid's book, *Facility Manager's Guide to Security: Protecting Your Assets*, provides the reader with an understanding of security and outlines the steps that are necessary to improve facility security. Beginning with the steps necessary to perform a risk assessment, the book offers the reader a clear cut understanding of determining a facility's assets. Complete with case histories and models the reader is walked through the risk assessment phase.

Anyone with facility management responsibilities will benefit from Reid's book. It discusses the various aspects of facility security including user access, electronic and mechanical access, building openings, etc.

From the most basic and elemental aspects of physical security—locks, alarms, building openings—to more advanced security measures like intrusion detection and CCTV, the reader is guided through the decision making process.

The step-by-step, easy to read format provides options for the book's readers. Facility managers, as well as those persons charged with the responsibility for a building's security, are given the "tools" for determining the most appropriate and viable options.

Preface

This book is written for people who are responsible for facility security. It tells the manager everything he/she needs to know about how to build and manage a security system. It talks about layout of the facility in order to provide secure facilities for workers, plants, processes and products. It explains how to perform a risk assessment. It explains security for different types of facilities like commercial property, schools, hotels, factories, warehouses and government buildings. It covers distances and separations for security, fencing, access control, key control systems, locks, hasps, gates, and the more sophisticated systems such as intrusion detection systems in burglar alarm systems, closed circuit television, infrared detectors, lighting, clear zones and electronic control systems. It addresses pass keys, mortise and cylinder locksets, master key, key control programs, thumbprint and retinal scan access control, voice print analysis, hand geometry, personnel badging, key pad, electronic key control, drug tests, security clearances and background checks. Finally it helps the facility manager to manage the security personnel, guards, rounds, shift work, training, procedure development and those other non-hardware elements of a security program.

Acknowledgments

One of the great things about the human spirit is that people can work together to create something better than individuals working alone. This spirit is fulfilled by the freedom of the American Experience. This book, *Facility Manager's Guide to Security: Protecting Your Assets,* could not have been done without the combined efforts of many people. I would like to thank the following people who have helped to make this book. First, John Orchard and Dale Bickenbach who looked at early versions of the work and provided valuable comments. Next, Larry Taylor, Sheriff of Benton County, Patricia Trainor of Moon Security Systems; Tom Koche of Desert Hills Video Security Systems; David Haneline of Scolari's Markets, Dave Rendleman of the Pasco Police Department and Anne Reed if the Pacific Northwest National Laboratory for helping with special applications and interviews. I would also like to thank Bill Phillips for his excellent work on Locksmithing and Warren Ryan of Washington Research for his help background checking. I especially want to thank Kevin Murray of Murray and Associates for assistance with a chapter on Electronic Surveillance and Countermeasures. All of these people work hard to help each other and to make the area around them more secure. By doing so, they help to make the world more secure. Thanks.

Also, there are many people who work in the security business who don't want to talk about it. This is understandable and I've tried hard not to weaken the systems, but to make them stronger.

If there are any mistakes, the author has to take it on the chin, so I'm sticking mine out.

Robert Reid 2004

Introduction

This book, *Facility Manager's Guide to Security: Protecting Your Assets*, provides facility managers and security personnel the tools they need to manage and improve facility security. It also provides the tools for upgrading a security system. It has been written for managers who want to find out how good their security is and if it needs improvement, how to improve it efficiently and cost effectively. It is a guide to help managers and facility staffs perform a risk assessment and determine if the present security is adequate. It provides guides for managers in the areas of perimeter security, separation distances, clear zones, fences and barricades. Also covered are locks, key control, electronic access and entry tracking, CCTV systems, audio surveillance defense, guards and guard management. It includes information on security programs, background checks, drug testing and badging employees. A chapter has been added about emergency response. It does not address computer security or viruses or worms or trojans. Those topics are for computer network security personnel.

It is a step-by-step guide to understanding security and it provides instructions for security improvement should it be necessary.

The book begins with an assessment of the assets that need to be protected, then shows how to perform a risk assessment against those assets. It provides the tools for determining how much security to invest in and how much it costs. Is it worth it to protect the asset? Probably, and this book tells how.

Included are case histories, models and samples to show exactly what is needed to perform a risk assessment and if security improvement is needed, it tells how to do it. It includes how to assess perimeter security, types of building construction, doors, walls and other openings and how to control and protect them. It includes detailed explanations of door hardware and control, locksets and key management. Also includes costs and replacement of door hardware and keys. It includes assessment of personnel doors and vehicle doors for commercial and industrial facilities.

The book goes further into access control by explaining how locks work and which ones are more secure than others and why. It explains how to assess the risk to a facility and whether or not to upgrade door hardware, how to do it and how much it will cost. The book has guides, illustrations and lessons learned to help make the best and most cost effective decisions for a facility.

The material includes all of the newer types of electronic hardware for door control, pass keys, electronic locks, card readers and many of the more recent electronic devices used by employees for access into buildings or into controlled spaces within buildings. It explains how these systems work and how to have them installed. It also provides information on the latest computer systems used to track personnel, use of card readers and how to provide access control for a facility using these new systems.

However, the book does not stop with getting in. It goes on to discuss intrusion detection devices and how they work. It also addresses other types of intrusion detection equipment like infrared sensors, ultrasonic listening devices, microwave transmitters, smart wire fences and many more types of applications.

Included are the many applications of closed circuit television systems (CCTV) and how they work, how they are installed and how much they cost. What type of cameras does a facility need to consider when buying? It includes samples of different types of CCTV cameras and what the limitations are.

It also covers the electronic applications of integrating these systems and the hardware and software to support them. How effective are they, how much do they cost and are they worth it?

This book addresses the power supplies for these integrated systems and how to make sure they are robust enough to support even the most difficult applications.

The book provides guides to badge personnel, how to set up a badging program and how long it takes to badge employees. What questions can legally be asked and where to store the information. The book also provides information on setting up a drug testing program. The book provides insights into management of security forces, guards and the trade-offs of hiring internal guard staff or contracting this service out. Lessons learned provide insight into the advantages and disadvantages of each type of program. The book also provides guides into setting up an emergency response program.

Facility Manager's Guide to Security: Protecting Your Assets is full of guides, charts, surveys and illustrations to make your job more cost effective and easier. Included are case histories and lessons learned that can be applied to every facility.

The first task begins with a Threat Assessment in Chapter 1. Enjoy your new security.

Chapter 1

Threats to Facilities

TYPES OF FACILITIES NEEDING PROTECTION

The first step in determining the status of a facility's security system is to determine the type and function of the facility. Different facilities face different threats depending upon what types of activities take place in the facility. This chapter discusses threats to various types of facilities and what security measures are commonly taken to mitigate the threats.

Not all facilities are alike, however, security at each facility must be tailored for that facility. The following tables are meant to be used as a guide.

Table 1-1 is a list of typical facilities and Table 1-2 is a list of common security measures taken at those typical facilities. In addition to an explanation of the terms in the tables, this chapter includes some typical case histories for facility upgrades. Finally, many facilities are a combination of types of facilities indicated the two tables. A facility may have a dam and a campsite in it or a resort may include a medical clinic on its grounds.

SECURITY SYSTEMS—A BREAKDOWN

The first step in physical security is separating the assets from the threats. This is often accomplished by separating the property where the assets are from the public. Most commonly this is done with fencing the perimeter and controlling access into or out of the site.

In some facilities it is neither practical nor feasible to separate the property. For example, restaurants or hotels rely upon customers coming and going for business and hence physical separation isn't feasible. So the next step is to control some of the

Figure 1-1. A typical hotel. A hotel is an example of a facility that needs physical security. (*Photo by Robert Reid, 2004*)

assets within the property by physical separation inside. This is usually done by separating areas within the facility and controlling access to the assets with secured doorways. Depending upon the sensitivity of the asset, the level of security for the doorway can be increased. Extra door security can be provided by locks, electronic surveillance or even the presence of armed guards. The sophistication of the door locks and the key control is another factor affecting the level of security that can be attained. In later chapters the book will explain the security possibilities for doors more extensively but for now the type of facility reflects what has been accepted in the industry.

As with doors, windows serve the occupants of a facility with light and environmental control. Many facilities have windows that do not open, and environmental control is provided by the facility air conditioning and heating systems, but older facilities have windows that can be opened. Another facet of windows is that they can be broken by an aggressor and access gained through the opening. So the presence of windows and the level of security for the facility can be affected by the size and type of windows.

A facility can have an extra area of protection by having a safe or a vault on the premises. Some facilities may have them and others will not.

A facility can be protected with what is called an Intrusion Detection System (IDS). Intrusion detection has become more sophisticated with the use of cameras, infrared scanners and sound measuring devices. However, Intrusion Detection Systems only notify the security force of intrusion, there is no notification of anything other than someone moving around within the protected area.

Another element used by facilities for security control is the Closed Circuit Television System (CCTV). Some facilities use a combined Intrusion Detection System and Closed Circuit Television System. In a different type of facility these systems aren't necessary or beneficial. For example, an Intrusion Detection System would not be practical in an all night restaurant or in a facility where there are people constantly working. A CCTV System might not be practical in a restaurant.

Facilities may benefit from enhanced lighting in the parking lot, especially if the employees are concerned about violence in the neighborhood. However, for a facility that does little at night and has very few people working outside of routine business hours, parking lot lighting might not be a good choice. Night lighting, also, affects the performance of Closed Circuit Television Cameras so there is a relationship between the CCTV system and the lighting systems.

Finally, to enhance security, a facility may need a security force. In some cases the security force may carry weapons, i.e. armed. Employees may need badges to separate staff from guests and finally, depending upon the work performed in the facility, it may be necessary to have some type of security background check performed on the employees and visitors.

These elements are all discussed more thoroughly in later chapters with specifics on how to determine whether these systems are effective and whether to modify them. The information presented here is for facility managers and risk assessment teams to get a feel for the facility and the types of security systems that may be required.

FACILITY FUNCTIONS AND THREATS

Table 1-1 provides some insight into the threats that may be encountered in different types of facilities. Some threats are low in one type of facility while that threat might be greater in a different facility.

Property Loss

A common threat in many facilities is property loss. Property loss includes loss that results from events like fires or floods. High winds can damage property and cause a loss. In addition people misplace things. The items may have been taken by another person or they may just be lost. A hotel is an example where property loss occurs. The facility has many transients, people who come and go, some who are accountable to the facility, and some who are not. In general property loss includes small items like luggage, parcels, handbags and laptop computers that are inadvertently left behind. Depending upon the location and type of facility, risk of property loss can be large or small.

Figure 1-2. A high school is a typical facility that needs security. (*Photo by Robert Reid, 2004*)

Table 1-1. Typical threats to facilities.

Facility Types	Property Loss	Theft	Violence	Unauthorized Access	Lost Persons	Environmental Hazards	Vandalism	Protests/Media Attention	Industrial Espionage
Airport	yes	yes	yes	yes	yes	possible	unlikely	unlikely	no
Amusement Park	yes	yes	yes	yes	yes	unlikely	yes	unlikely	no
Apartment Building	yes	yes	no	unlikely	unlikely	no	possible	no	no
Bridge	no	no	no	possible	no	no	possible	possible	no
Campground or Site	yes	yes	yes	yes	unlikely	no	yes	no	possible
Commerical Building	yes	yes	yes	yes	unlikely	possible	yes	yes	yes
Courts or Courtrooms	yes	unlikely	yes	yes	yes	unlikely	yes	yes	possible
Dam	yes	yes	unlikely	yes	unlikely	yes	yes	yes	yes
Dock/Harbor	yes	yes	yes	yes	yes	yes	yes	unlikely	yes
Factory	yes	yes	yes	yes	unlikely	yes	yes	unlikely	yes
Food Processing Facility	yes	yes	yes	yes	yes	unlikely	yes	possible	yes
Gambling Casino	yes	yes	yes	yes	yes	yes	yes	possible	yes
Grocery Store	yes	yes	yes	yes	yes	yes	yes	possible	yes
Hospital	yes	yes	yes	yes	yes	yes	yes	unlikely	possible
Hotel	no	yes	yes	yes	no	no	yes	yes	no
Jail or Prison	yes	no	yes	no	yes	yes	yes	unlikely	yes
Medical Clinic	yes	yes	unlikely	yes	yes	yes	yes	unlikely	yes
Military Base	yes	yes	yes	yes	unlikely	yes	yes	yes	yes
Mining Facility	yes	yes	yes	yes	unlikely	no	yes	yes	no
Monument	no	yes	yes	no	yes	no	yes	yes	no
Parking Lot	yes	yes	yes	no	no	unlikely	yes	unlikely	no
Post Office	yes	yes	yes	unlikely	no	possible	yes	yes	yes
Powerplant	yes	yes	yes	yes	yes	yes	unlikely	yes	no
Rail Station	yes	yes	yes	yes	no	possible	yes	no	possible
Residence	yes	yes	yes	yes	yes	yes	yes	unlikely	yes
Resort	yes	yes	yes	yes	no	unlikely	yes	possible	no
Resturant	yes	yes	yes	no	no	yes	yes	unlikely	yes
Retail Store	yes	yes	yes	yes	yes	unlikely	yes	yes	yes
School	yes	yes	yes	yes	yes	yes	yes	yes	no
Stadium	yes	yes	yes	yes	no	yes	yes	no	no
Tunnel	no	no	no	yes	yes	yes	yes	no	no
Univiersities	yes	yes	yes	yes	yes	yes	yes	yes	yes
Warehouse	yes	yes	unlikely	yes	no	yes	yes	unlikely	yes

Theft

In addition to property loss, theft is another threat to a facility. However, in a facility such as an office where there are few guests and the staff knows each other the chances of a theft are lower than in a facility where there is a lot of transient traffic. For a factory that makes large items like, for example, air compressors, the odds of a theft of one of these units is smaller than the theft of employee personal items or hand tools. Facilities that fabricate large items have to be concerned about the theft of component elements. Theft can also include larger items that require heavy equipment to haul. Military bases for example have had examples of thieves attempting to steal a complete jet engine using a large truck.

Violence

Facilities must continue to be vigilant against the treat of violence in the workplace. Although most violence takes place in the home where the assailants know each other, there are occasional random acts of violence. Violence can be anything from a fight breaking out between employees, to sexual assault, to murder and terrorism. A facility cannot be made completely safe from violence but much can be done to prevent it. Judicial courts typically search persons entering the facility to make sure they are not carrying weapons. That way, the violence they could perform is minimized. Courts also use closed circuit television systems to record actions of people in the courtrooms and hallways. Schools have installed CCTV systems, which have helped to curb violence in these facilities.

Unauthorized Access

At one time or another all facilities face the threat of unauthorized access by a person or persons. Unauthorized access is defined as a person being in an area of a facility where they are not allowed. In many facilities simply posting signs saying "Authorized Personnel Only" is satisfactory, but in other facilities, more positive steps are taken including door locks and security guards. In a few types of facilities unauthorized access is more difficult to control. For example, a hospital or a medical facility is a facility where patients, staff, friends and family are all users in the facility. In a facility such as this, it is difficult to prevent unauthorized access without some type of control.

Lost Persons

In addition to unauthorized access some types of facilities are more subject to lost persons than others. The lost or disoriented person can be a patient in a medical facility or a child in an amusement park. If there is a lost person, then there will be additional people looking for the lost person so the threat of a lost person in a facility can lead to unauthorized access into areas where intruders should not be allowed. Mitigating factors for lost persons include security forces and signs telling persons where they are and giving guidance to where they might need to go.

In other facilities, lost persons are not likely. For example in a secured facility with a trained staff where visitors are not allowed, the chances of a lost person are reduced. At another type of facility, e.g. a county or state fair, the possibility of a lost person increases. A lost person is a threat to a facility's security because resources must be used to look for the lost person and those looking may have interests other than finding the lost person.

Environmental Hazards

The threat to a facility from environmental hazards is one that is difficult to define and difficult to cope with. Environmental hazards would include acts of sabotage through nuclear, biological or chemical materials. Facilities that conduct business that are highly visible to protest groups or terrorist groups are subject to the threat of chemical or biological poisoning of air, water or food. A large public facility that is perceived to affect the environment is subject to this type of threat which would include chemical sprayed into the ventilation system or a hazardous material transmitted through the mail. Courts are an example of a facility targeted in the past. However, other facilities are less likely to be subjected to this threat. For example, an apartment building, where there are many occupants, there is no group where attention is likely to be focused. In this type of facility it is less likely that this type of threat would be attempted. Structures that are primarily outside such as public monuments, highway bridges or campsites make it difficult for the environmental hazard to be effective.

Table 1-2. Typical security measures to counter security threats for common facilities.

Facility Type	Gate	Fencing	Vehicle Barriers	Gate Guards	Door Locks	Extra Door Secuirity	Keys and Key Control
Airport	Yes	Yes	Yes	Yes	Yes	Yes	Yes
Amusement Park	Yes	Yes	Yes	Yes	Yes	No	Yes
Apartment Building	Possible	Yes	No	Possible	Yes	No	Yes
Bridge	No	No	No	No	Yes	No	Yes
Campground or Site	Yes	Yes	Yes	Yes	No	No	Possible
Commercial Building	Possible	Possible	No	No	Yes	No	Yes
Courts or Courtrooms	No	Yes	Yes	Yes	Yes	Yes	Yes
Dam	Yes	Yes	Yes	Yes	Yes	Yes	Yes
Dock/Harbor	Yes	Yes	Yes	Yes	Yes	Possible	Yes
Factory	Possible	Yes	Possible	Possible	Yes	Possible	Yes
Food Processing Facility	Yes	Yes	No	Possible	Yes	No	Yes
Gambling Casino	No	No	No	No	Yes	Yes	Yes
Grocery Store	No	No	No	No	Yes	No	Yes
Hospital	No	No	Possible	Possible	Yes	Possible	Yes
Hotel	No	Possible	Possible	Possible	Yes	No	Yes
Jail or Prison	Yes	Yes	Yes	Yes	Yes	Yes	Yes
Medical Clinic	No	No	No	No	Yes	No	Yes
Military Base	Yes	Yes	Yes	Yes	Yes	Yes	Yes
Mining Facility	Yes	Yes	Possible	No	Yes	No	Yes
Monument	Possible	Possible	Possible	Possible	No	No	Yes
Parking Lot	Possible	Possible	Possible	Possible	No	No	No
Post Office	No	Yes	Yes	No	Yes	No	Yes
Powerplant	Yes	Yes	Yes	Yes	Yes	Possible	Yes
Rail Station	No	No	No	No	Yes	No	Yes
Residence	Possible	Possible	No	Possible	Yes	No	No
Resort	Possible	Possible	No	Possible	Yes	No	Yes
Resturant	No	No	No	No	Yes	No	Yes
Retail Store	No	No	No	No	Yes	No	Yes
School	No	Yes	Possible	No	Yes	No	Yes
Stadium	Yes	Yes	Yes	Yes	Yes	No	Yes
Tunnel	No	No	No	No	Yes	No	Yes
Univiersities	Possible	Yes	Yes	Possible	Yes	Possible	Yes
Warehouse	No	No	No	No	Yes	No	Yes

CASE STUDY

Governor Receives Radioactive Jam in the Mail

In 1990 maverick scientist Norm Buske, a protester opposed to the continuing nuclear waste spills at the Hanford Reservation in southeast Washington sneaked onto the facility and harvested a large quantity of mulberries growing along the river where environmental nuclear waste was thought to have spread. The individual made jam from the berries and shipped the jam as a gift to then Washington State governor Booth Gardner with a hazard note attached that the jam might be radioactive. It was.

Table 1-2. (*Continued*)

Window Locks	Security Guards	Badges for Staff	Door Access Control	Safes/Vaults	Intrustion Detection	Closed Circuit TV	Enhanced/Night Lighting	Employee Background Checks
No	Yes	Yes	Yes	Yes	No	Yes	Yes	Yes
Yes	Yes	Possible	Yes	Yes	No	Yes	Yes	Possible
Yes	Possible	No	Yes	No	No	Possible	Yes	Possible
No	No	Yes	Yes	No	No	Yes	Yes	Yes
No	Possible	No	No	No	No	No	No	No
Yes	Possible	Possible	Possible	Possible	Possible	Possible	Yes	Possible
Yes	Yes	Yes	Yes	Possible	Possible	Yes	Yes	Yes
Yes	Yes	Yes	Yes	No	Yes	Yes	Yes	Yes
No	Yes	Yes	Yes	Possible	Possible	Yes	Yes	Possible
Yes	Possible	Possible	Yes	Possible	Possible	Possible	Yes	Possible
Yes	No	Possible	No	No	No	Possible	Possible	Possible
No	Yes	No	Yes	Yes	Yes	Yes	Yes	Yes
No	No	No	No	Yes	No	Yes	Yes	Possible
Yes	Possible	Yes	Yes	Yes	No	Possible	Yes	Yes
Yes	Possible	No	Yes	Yes	No	Yes	Yes	Possible
Yes	Yes	Yes	Yes	Possible	No	Yes	Yes	Yes
Yes	No		Yes	Possible	Yes	No	Possible	Possible
Yes	Yes	Yes	Yes	Yes	Yes	Yes	Yes	Yes
Yes	Possible	No	No	No	No	Possible	Yes	Possible
No	Possible	Possible	Yes	No	Possible	Possible	Yes	Possible
No	Possible	No	No	No	No	Possible	Possible	No
Yes	Possible	No	No	Possible	Yes	Yes	Yes	Yes
Yes	Possible	Yes	Yes	No	Possible	Yes	Yes	Yes
Yes	Possible	Yes	Possible	No	No	Possible	Yes	Possible
Yes	No	No	No	Possible	Possible	Possible	No	No
Yes	Possible	Possible	Possible	Yes	Possible	Yes	Yes	Possible
Yes	No	No	No	Possible	No	No	Yes	No
Yes	Possible	No	No	Possible	Possible	Yes	Yes	Possible
Yes	Possible	Yes	No	No	Possible	Yes	Possible	Yes
No	Yes	No	No	Possible	No	Possible	Yes	No
No	No	No	No	No	No	Possible	Yes	Possible
Yes	Possible	Yes	No	Yes	Possible	Possible	Yes	Yes
Yes	Possible	No	No	Possible	Yes	Yes	Possible	No

The point in this case is that a person, in an attempt to protest, created a potential hazard to the governor and a considerable media event. Hanford is still trying to clean up the radioactive waste and will be for many years in the future.

Vandalism

Many facilities are subject to the threat of vandalism including schools, other public buildings and businesses. Vandalism is generally not a sophisticated threat, it is the result of persons who lack intelligence and discipline. Vandalism can take the form of spray painted graffiti, damage to trees and shrubbery, broken windows, broken light fixtures and other minor property damage.

CASE STUDY
The Tipping Point for the New York Subway System

In the late 1980s the New York subway system had a serious vandalism problem with graffiti and trash in the subway cars. The problem was that the penalty for these acts was small and the City did not have the budget for cleaning it up. However when Bernard Goetz was assaulted on the subway in 1984 and he retaliated violently by shooting his 4 assailants, the City decided to do something about cleaning up the subway. Graffiti was no longer tolerated and cars were immediately repainted if graffiti was present. Trash was cleaned up. The City determined that property that looked as if no one cared about it would be subject to more vandalism than otherwise. This theory is called "The Broken Window Theory" after a technical paper written by Henry G. Cisneros in 1995. The meaning being that an empty house with a broken window will soon have more broken windows because there is no penalty to the vandals. The New York City subway system was cleaned up. People who jumped toll gates were stopped and challenged. Minor infractions were no longer tolerated. As a result the New York subway system became much safer and is today a model of modern transportation.

Source: Malcolm Gladwell. *The Tipping Point: How Little Things Can Make A Big Difference,* Little, Brown and Company, Copyright 2000.

Protests or Extra Media Attention

Any facility can become a target for a protest but some facilities are more likely to be protested than others. Nuclear power plants and other large industrial complexes have been targeted in the past. Recently the large chain of Wal-Mart Stores has been targeted by protesters wanting to resist the impacts a large commercial retail facility brings into a neighborhood or town. While often the protest group is small, under 25 persons, they receive a lot of media attention which can affect business, the goal of the protest group. Some facilities are more likely to see protests and media attention than others. Airports are unlikely to be protested, nor established resorts. Dealing with the media is addressed in Chapter 16: Emergency Response.

Espionage

Any government, court or business is likely to be threatened by espionage. Generally invasion of privacy is illegal but public scrutiny may or may not be criminal. For example is searching the papers in the wastebasket of the town hall meeting by a reporter illegal? Business may want to find out how another business is doing and as long as public domain information is used, this is not illegal. Of course government's, desire to keep military plans and policies secret but in a similar way, a business may want to keep a bid for a new project secret. Espionage can be an external threat, from persons outside the system, or it can be an internal threat, from existing employees. Espionage is unlikely in large pubic facilities. If the facility is public what is the secret? Also, espionage is unlikely in public monuments, stadiums and apartment buildings.

Summary

Table 1-1 and Table 1-2 show the threats and security measures for many typical facilities. Some facilities may encompass more than one type. As noted, the information in the tables is only for guidance but each facility should perform its own assessment. Issues can be regional or national in scope so an unlikely threat at one type of facility in one region or state may be more likely in another region or state. Values and laws vary from region to region and internationally laws vary widely due to custom, tradition and religion. What is illegal in one country may be considered good business practice in another. Hence it is important to understand local values as well in applying risk and security measures.

CASE STUDY
Hospital Security Upgrade—
Newton-Wellesley Hospital in Newton, Mass.
The Nelson-Wellesley hospital in Newton, Mass., is a 6-floor

233-patient-bed facility with 1400 employees and another 450 doctors who have visiting access. In order to increase security the hospital upgraded its security system in 2002/3. The primary reason for the security upgrade was to improve access control throughout the facility and reduced the number of security incidents. The decision was based upon the results of a risk assessment performed in 2000 that had identified vulnerability in the medical library where research materials were kept, in the medical records area, in maternal and child health areas, and the numerous access doors with many keys and a cumbersome key control program.

After an assessment and with the large number of employees the key control system was determined ineffective so the facility installed 100 proximity sensors throughout the campus and issued proximity cards to staff. (More information of proximity card readers and proximity card is presented in Chapter 8 which covers electronic locks in more detail.)

In addition to additional door control, the facility upgraded the closed circuit television system (CCTV) at the same time adding 40 high-speed dome cameras linked to three digital video recorders to maintain surveillance over the main entrances, interior areas and parking lots.

Approximately 2,500 proximity cards were issued well in advance of implementation of the new lock process and the security department worked with the staff to educate them on use of the new systems and ease any concerns that the new system would adversely affect medical or hospital support staff's ability to provide patient care.

Surveillance Specialties Ltd., of North Andover, Mass., installed the new systems. The contractors had to cut access panels into the walls, install door contacts, the card readers and other motion sensors to make the new access control system compatible with local fire codes and with the requirements of the Joint Commission for Accreditation of Heath Organizations (JCAHO). The JCAHO is a group that assists with hospital standardization nationwide.

Because the work was performed in a working hospital, close coordination was required between the contractors installing the systems and hospital staff. If creating dust or noise was a problem, the work was rescheduled for a better time.

The first phase was completed in December of 2003 reducing security incidents on campus, strengthening the hospital's emergency response capability, and bringing the facility in to full compliance with the Health Insurance Portability and Accountability Acts (HIPAA) and the Joint Commission for Accreditation of Heath Care Organizations (JCAHO).

This information is presented with Permission from Newton-Wellesley Hospital.

CASE STUDY
Resort Security Upgrade

The Caribe Hilton Resort in San Juan, Puerto Rico, decided it was time up upgrade the Camera Surveillance System when the hotel was renovated in 2000. The facility consists of three buildings near the beachfront. The first building is the main hotel, originally built in 1949, the second is a garden wing adjacent to the main building and finally a 20 story tower overlooks the ocean. The resort has 646 rooms and includes pools, a playground, a shopping arcade, restaurants and a health club. The facility also has 60,000 square feet of meeting space for conferences.

The decision to upgrade the surveillance system was based upon the risk of theft of private property, mostly laptop computers and personal digital assistants. Also, of concern, was the outdoor lobby; which was sometimes slippery after a brief rain shower. Finally, there had been a problem with vendors damaging elevator doors and the facility thought that digital cameras would help establish accountability and hopefully encourage vendors to reduce damage.

The existing system had 16 cameras which were not enough to cover the property and the backup was video tape recorders which were difficult to search when an incident occurred. The budget for the upgrade was $70,000 dollars which would cover the addition of another 36 color cameras with pan, tilt and zoom capability. Three digital video multiplex recorders (DVMRs) would store the data from the cameras. The project included installing the necessary wire and cable for the new equipment.

Unfortunately the original building was concrete, making installation of new wire and cable difficult. This problem was offset by spare telephone cables in the facility comprised of unshielded twisted pair. The security system was able to use some of these telephone spares and save costly new wiring runs.

The new equipment was chosen for its ease of integration and compatibility with the old equipment. The new DVMRs were capable of feeding information to multiple security TV monitors. The project included an upgrade to the security office that required a new central station with 5 monitors, the three DVMRs and two keyboards to control the equipment. In addition, the security station had to house a telephone switchboard and the fire alarm monitoring system. As a result it was necessary to install a custom made console for the equipment. Tinted windows were installed in the security station to prevent a casual visitor from interfering with the monitoring by the security force.

For some outside cameras it was necessary to install additional power cable in conduit. This work had to be performed by a separate contractor and close coordination between contractors was a challenge. The power supplies were connected to the hotel uninterruptible power supply system.

Housekeeping was also able to use the information from the CCTV monitors to maintain the facility, preventing slipping on wet floors. The number of thefts of personal equipment went down significantly. The value to the hotel has more than paid back the initial investment in terms of customer satisfaction and loyalty. *This article originally appeared in Security Management Magazine in July 2003 and was written by Joanne Harris who prepared this information for Dedicated Micros. It is reprinted here with permission from Dedicated Micros™.*

———————

Chapter 2
Performing Risk Assessments

BEGIN WITH AN ACCURATE AND
UP-TO-DATE RISK ASSESSMENT

The next step to understanding the status of security in a facility is to conduct a risk assessment. A risk assessment is a systematic analysis of the facilities assets and an assessment of the vulnerability of those assets from different types of threats. By performing a risk assessment facility managers can understand what is valuable and what the threats are. Then the facility can assign a priority to protect the most vulnerable assets.

UNDERSTANDING
THE FACILITIES' ASSETS

Security begins with a thorough knowledge of the assets and protection of those assets begins with knowing what to protect. After a proper risk assessment it can be seen which assets are the most valuable and which are the most vulnerable. Security then takes what resources are available and assigns them to protect the most important elements first.

For example, assets to be protected include the property, inventory, raw materials, product, the production line. Another important asset includes the people in the facility whether they are employees, customers or guests. Staff knowledge is a valuable asset that cannot be insured.

KNOWLEDGE

Good security begins with knowledge of the assets and a large part of this knowledge is information. Who are the custom-

Steps to Perform a Security Risk Assessment

Step 1. Decide upon a risk assessment team.
  Size of Team? _____
  Members should be from management and have good working knowledge of facility assets.

Step 2. Review and list the assets.
  People (or labor) assets (how many people and their value?)
    1. _____   2. _____   3. _____
    Continue on until all people assets are listed.
  Capital Assets (Cash, Bonds, Real Property.)
    1. _____   2. _____   3. _____
    Continue on until all capital assets are listed.
  Material Assets (Personal Property, Inventory, Tools, Product)
    1. _____   2. _____   3. _____
    Continue on until all material assets are listed.

Step 3. Determine the value of the assets. What would happen if the facility lost this asset?
  This value is based upon:
    book value? _____
    replacement value? _____
    loss of production? _____
    loss of revenue? _____
    liability?_____
    penalties? _____
    other method? _____

Step 4. Determine the threat to the asset.
  Who are the aggressors?
  Does the asset benefit the aggressor?
  Who knows about the assets? Are they public?
  Have there been incidents involving aggressors in the past?
  Have there been incidents nearby with aggressors in the past?
  Is there a potential for future incidents with aggressors?
  Is the asset accessible?
  Is there law enforcement or other security that acts as a deterrent?

Step 5. Determine the severity of the risk to the facility by checking the threats against the assets.

Step 6. List the assets needing protection in order from the highest (needing most protection) to the lowest (needing less protection.)

ers and what they owe might be more important than the cashbox so a risk assessment requires tabulation of what the company has that is valuable.

A problem that immediately arises once the assets are being inventoried is how to protect the information of what the assets are. A person with a lot of cash does not want to advertise it. In fact, one element of security is knowing what the asset is and then preventing others from knowing what the asset is or where it is. Obviously the fewer people who know what the most sacred assets are the easier it is to protect them.

Therefore, the first decision to be made after the decision to list the assets is: Who is going to see this information?

THE RISK ASSESSMENT TEAM

In order to prepare a proper risk assessment, it is important to determine who will list the assets. This team can be small or large depending upon who ultimately pays for the security. In a small company, a small team or even the owner can decide what the assets are. The important point is to understand that with more team members more assets are going to be added to the list of assets to be protected. The manager who builds the team should recognize that the job multiplies according to the dynamics of the group—the larger the group, the larger the list.

Nevertheless, determination of assets performed by more than one person is better determination of assets by a single individual, because a single individual is likely to miss a critical component of the assets.

The team members who perform the risk assessment should include people who are familiar with the assets of the facility. Therefore the risk assessments team should include the manager or executive in charge of each element. For example in a business with multiple offices the risk assessment team should include managers of each facility. For a factory with more than one operational element each manager of the element should be included. In addition it is important to have a representative from shipping and receiving on the team, as well as, the manager of security. Depending upon the information held by the personnel staff it may be necessary to include a member from human relations on the risk assessment team.

The scope of work of the team should be guided. Usually, the first step is to determine and list the assets and then determine what the threats are. But some facilities perform this function in the opposite way, first seeing what the threats are and then determining what the assets are. Either way is satisfactory.

Another important point when evaluating assets is the relative time sensitivity of some types of information. For example in case of a company wanting to protect its bid it only needs to be protected between the time the bid is prepared and the contract is awarded. If it is a public bid, it only needs to be protected between the time when the bid is prepared and the bids are publicly opened. Not only are assets valuable because of what they are, but they are also valuable because of when.

KNOW YOUR ASSETS

Good security consists of knowing what the assets are and then determining what their values are and deciding if the asset is worth protecting. Assets can take many forms but the main test is to determine if it has value. The risk assessment team should consider the following types of assets at the beginning of the process.

For a business to survive it requires personnel or labor, capital or money, and material and knowledge.

LABOR

One of the main assets is the labor force of the facility. It doesn't matter whether the company provides a service or product, the people in the facility and their knowledge can be one of the most valuable company assets. Every facility should understand who the employees are and what they do.

CAPITAL

Next on the list of assets to be evaluated is the capital available to the company. This would include cash on hand, including

In one small company all of the corporate officers enjoyed flying and one of the corporate officers owned a light plane that the officers would use to fly to and from business meetings throughout the state. One morning on the way to conduct a presentation the plane crashed, killing 5 of the 6 corporate officers. The company survived, but only barely. The loss of corporate knowledge and the ability to perform work was severely crippled. The company was set back quite a few years.

This is one situation that should be considered when the value of labor is assessed in the risk assessment. Many companies do not allow more than two corporate officers on the same airplane flight for this reason.

the source of any access to short and long term loans. If resources are held in a bank or other financial institution, the risk team can rely upon the risk mitigation resources of the institution for that portion of the risk assessment.

MATERIAL

Next in the risk assessment, the team should consider any raw materials, partially completed product or the final product. All three are assets to be included in the risk assessment. Also included in this category includes tools used by the facility to make the products. In some special applications, the tools can be one of the most important and costly assets.

KNOWLEDGE

Finally, in the risk assessment, the team should consider the value of corporate knowledge. Special contracts, agreements, or contacts would be included in this part of the risk assessment. Time sensitive information should be considered. However, since the results of the risk assessment are to outlast any one event, the time sensitivity should be put into the risk assessment in such a way that the knowledge is encompassed. For example, a companies' bid on a project is a time sensitive asset. A single bid would not be listed in the risk assessment; but, in general, sealed bids would be one item to include in the risk assessment. Other knowledge for consideration by the risk assessment team would include sensitive payroll or personnel information, names of important contacts, plans for special projects and information that is sensitive but not of a physical nature.

A NOTE ABOUT COMPUTER SECURITY

Since much of the knowledge is contained in computers it is important to protect computer networks and systems from intrusion by unauthorized personnel. Since this book is not about computer security, but about physical security, those needing information on computer security should check into other books on that subject. A couple of good references for computer security are:

Hack Attacks: How to Conduct Your Own Security Audit by John Chirillo 2003 Wiley Publishing, Inc., Indianapolis, IN.

Hackers' Challenge 2: Test Your Network Security and Forensic Skills by Mike Shiffman et al. 2003, McGraw-Hill/Osborne, Berkley, CA.

However, much computer security can be accomplished by maintaining a firewall, using up to date software and keeping anti-virus software current. These three simple steps provide a great deal of protection for computer systems. Microsoft® maintains up to date software products on its web site at http://windowsupdate.microsoft.com/. This web site maintains current updated versions of software and is downloadable free from the internet.

Finally, choose a good anti-virus software program and keep it up to date. Anti-virus programs are available on the internet and some basic programs are provided at low cost to the user.

THREATS

Once the assets are known and understood, the next step in the risk assessment is to evaluate the threats to the assets. A threat assessment is one piece of the overall risk assessment. From a security standpoint threats to assets are generally man-made, while in an overall risk assessment, other events, such as natural disasters can be included.

Natural Threats

Natural threats to a facility's assets include disasters such as fire, flood, tornadoes, hurricanes and earthquakes. While these natural events cannot be prevented, they are well understood and many government resources have been committed to determining and preparing for these catastrophic events.

The National Weather Service tracks weather patterns and provides local news media of impending adverse weather conditions like lighting storms, intense rain, tornadoes and hurricanes.

The United States Geological Survey provides earthquake data.

Depending upon size and jurisdiction, local regional and national agencies provide advance warnings of fires.

Physical security measures cannot provide much to prevent these natural disasters but proper preparation by the facility and an awareness of the risks posed by such disasters can minimize damages from these events.

Another category of natural threats to facilities can be infestation or rot by termites or animals. Also, some forms of mold or rot can threaten a facility or the facility's occupants. These threats are countered by pest and termite extermination companies.

THREATS BY PERSONS

Physical security can prevent and mitigate the treats to facilities assets from intruders and aggressors. To address the threats accurately it is important to understand the nature of the aggressor and their ability to carry out the threat. Physical threats to assets by persons can be broken down into three categories.

Threats from Criminals

Criminals are persons or groups who purposely intend to carry out a threat. Criminals can be unsophisticated or sophisticated and can sometimes be organized into groups who perform criminal activities. The criminal category includes vandals. In a broad and general sense, criminals seek to take or damage the assets for their own gain or satisfaction. Threats from criminals include theft, burglary, vandalism, sabotage, arson, and embezzlement. Threats to persons include workplace violence, assault, battery, rape, kidnapping and murder.

Threats from Terrorists

Terrorists differ from criminals, in that their intent is to disrupt operations and destroy the assets. Terrorists can be individuals or part of a group and some terrorists are sponsored by other governments. If terrorists are from another country, they may be considered heroes in that country. Therefore a different approach is required to deal with terrorists. Threats from terrorists include bomb threats, threats conducted with the use of weapons, use of weapons of mass destruction, violence, sabotage, arson, kidnapping and murder.

A Third Category of Aggressors

Finally, there is a third category of aggressors who are a threat to assets. This third category includes activists whose protests pose a threat to a facility in the form activities designed to prevent work or halt production. Activists who picket a plant site fall into this category. While not actually breaking any laws, their presence can be a threat to persons in the facility or to assets within the facility.

The Activist/Protestor category of threat aggressor includes the extreme activist. An extreme activist is a person or group of persons whose protest is so violent they are willing to perform criminal acts. This would include violence or property damage to facilities assets.

The analysis of potential threats needs to be based upon data. Obtaining this type of data is not difficult; however it does require some effort. For the purpose of conducting a threat assessment, one data source comes from the Uniform Crime Report maintained by the US Federal Bureau of Investigation in the Department of Justice. The Uniform Crime Report database provides users with statistical numbers of major crimes in major cit-

The U.S. Army Technical Manual for Risk Analysis is AR190-51. In their manual they break down the aggressors into the following 8 categories.
  Unsophisticated Criminals
  Sophisticated Criminals
  Organized Criminal Groups
  Vandals/Activists
  Extremist Protest Groups
  Terrorists within the Continental United States
  Terrorists outside of the United States
  Paramilitary Groups outside the United States

ies and tracks the trends in crime either up or down. Another report, the National Crime Victimization Survey also presents data on crime; however the National Crime Victimization Survey does not indicate crimes by geographic area.

The risk analysis team can determine the likelihood of a threat by using the crime data in the Uniform Crime Reports. For example, in the June 2003 Uniform Crime Report for Tulsa, Oklahoma in the period from January to June the following crimes were reported to law enforcement and to the FBI.

Tulsa, OK, January to June 2003

	Number of Crimes	Rate per 100,000 People
Violent Crimes	2078	530
Murder	28	7
Rape	125	31
Robbery	428	109
Assault	1497	381
Property Crimes	13,625	3476
Burglary	3343	853
Larceny/Theft	8601	2194
Stolen Cars	1681	428
Arson	133	33

What these data allow the risk assessment team to do is determine the relative threat to the installation from these types of crimes. The approximate population of Tulsa is 391,908.

In another example, Detroit, Michigan, is shown in the Uniform Crime Report for the period of January to June of 2003 as follows:

Detroit, MI, January to June 2003

	Number of Crimes	*Rate per 100,000 People*
Violent Crimes	8,833	954
Murder	156	17
Rape	282	30
Robbery	2,805	303
Assault	5,600	605
Property Crimes	30,120	3,256
Burglary	6,363	687
Larceny/Theft	12,092	1,307
Stolen Cars	11,665	1,261
Arson	—No Figures Reported—	

Again the population of Detroit is higher than Tulsa, so it is necessary for the risk assessment team to take this population difference into account. Detroit's population is 925,051.

Finally, a third city is Los Angeles, California. The Uniform Crime Report for this city reveals:

Los Angeles, California, January to June 2003

	Number of Crimes	*Rate per 100,000 People*
Violent Crimes	24,978	657
Murder	258	7
Rape	660	17
Robbery	8,512	224
Assault	15,548	409
Property Crimes	68,457	1,801
Burglary	12,424	327
Larceny/Theft	39,085	1,028
Stolen Cars	16,948	446
Arson	1,124	30

The population of Los Angeles is 3,798,981.

With this information the risk assessment team can determine the relative threats to the facility. For example larceny/theft for the three areas is (per 100,000:)

Tulsa	2,194
Detroit	1,307
Los Angeles	1,028

And as a result the risk assessment team can determine the threat from theft/larceny is more likely in Tulsa than in Los Angeles by almost a factor of two to one.

Another good source to check for the threat assessment is local law enforcement. Local data is more up to date and may prove more useful. In some areas the local law enforcement data may not be available or accurate. These are only initial threat data and are useful to determine the number and type of aggressors.

The above analysis was prepared using data from the US Government. For areas outside of the United States, the United Nations has the International Crime Victim Survey which includes criminal data from some other countries. Because of the variations in criminal definitions, international data cannot be directly compared to the US data. However, the UN data can be used to analyze threats and determine aggressors. Again local law enforcement may have better data depending upon the sophistication of the country.

With threat data from Uniform Crime Reports, United Nations data or local law enforcement and with an understanding of the assets and their values there are a few more steps required before the risk assessment can be completed.

VISIBILITY

Is the asset visible to the aggressor? Is the asset in plain sight or is the asset kept secured in an existing facility? Determination of the visibility of the asset to the aggressor is the next step. If the asset has low visibility or is unlikely known to the aggressor then

Much crime information about a facility's place of business can be located on the internet. The three cities, below, show the crime tabulation from 2001 for Tulsa, Oklahoma, Detroit, Michigan and Los Angeles, California, in 2001. The source of this information was www.city-data.com.

Crime in Tulsa (2001):
- 34 murders (8.7 per 100,000)
- 256 rapes (65.1 per 100,000)
- 776 robberies (197.4 per 100,000)
- 3,481 assaults (885.6 per 100,000)
- 5,863 burglaries (1491.7 per 100,000)
- 15,308 larceny counts (3894.7 per 100,000)
- 3,636 auto thefts (925.1 per 100,000)
- City-data.com crime index = 605.3 (higher means more crime, US average = 330.6)

Crime in Detroit (2001):
- 395 murders (41.5 per 100,000)
- 652 rapes (68.5 per 100,000)
- 7,096 robberies (746.0 per 100,000)
- 12,804 assaults (1346.0 per 100,000)
- 15,096 burglaries (1586.9 per 100,000)
- 29,613 larceny counts (3113.0 per 100,000)
- 24,537 auto thefts (2579.4 per 100,000)
- City-data.com crime index = 1003.9 (higher means more crime, US average = 330.6)

Crime in Los Angeles (2001):
- 588 murders (15.9 per 100,000)
- 1,409 rapes (38.1 per 100,000)
- 17,166 robberies (464.6 per 100,000)
- 33,080 assaults (895.3 per 100,000)
- 25,695 burglaries (695.4 per 100,000)
- 79,521 larceny counts (2152.2 per 100,000)
- 31,819 auto thefts (861.2 per 100,000)
- City-data.com crime index = 553.3 (higher means more crime, US average = 330.6)

the risk is lower. However for a different type of aggressor, the visibility may be higher. A good example would be the difference between the sophisticated criminal and the unsophisticated criminal. An asset may be known to a sophisticated criminal where it is not known to an unsophisticated criminal.

USEFULNESS

Is the asset useful to the aggressor? If the asset has immediate value then there is greater risk to the asset than if it has little value. The asset may have value as cash, where it can be sold or it may have value to the aggressor because he intends to use it. Many cars are stolen by a petty thief for joyriding, and they are found abandoned a few hours or days later. But another type of aggressor might cannibalize the car and attempt to sell the parts individually or on the Internet.

Finally, the aggressor might perceive that the asset has usefulness for its publicity value. What could happen to the company if the assets were taken and the public knew they were taken as opposed to few persons giving significance to the loss of the asset? The crime of tampering with medications in order to depress the stock value of the company comes to mind.

AVAILABILITY

Is the asset available to the aggressor? Would it be easier for the aggressor to take someone else's asset? Or would the aggressor try to obtain the asset because it is the only one available?

ACCESSIBLE

Is the asset accessible? What measures are already in place to make the asset difficult to access? Is the asset inside or outside? Is the area fenced or secured? Are there current systems being used to protect the asset like lighting, locks, accountability, guards, intrusion detection or alarm systems? Is there more than one layer of protection? This element will be discussed more thoroughly in Chapter 3 but the risk assessment is based upon what protection there is for the asset now.

LOCAL OR NEARBY INCIDENTS IN THE PAST

Have there been recent incidents in the area in recent times? The fact that these incidents have occurred in the past and nearby may indicate more than one aggressor or that the community has tolerated these incidents in the recent past. In addition, the aggressor may be trying to change the behavior of the facility by threatening the assets. It would be up to the risk assessment team to determine the relative importance of recent local or nearby incidents.

POTENTIAL FOR FUTURE INCIDENTS

What is the potential for future incidents? This element is subjective and difficult for the team to manage. However the team can assess whether the incident is probable, likely or unlikely.

DETERRENCE

Next on the risk assessment list is the likelihood of deterrence as perceived by some aggressors. Is there current protection to indicate that the chances to seize the asset are easy or difficult? Is the asset guarded, protected, secured? For example, if the asset sits out on the loading dock and the area is not secured, it may be easy for the aggressor to seize the asset. If it is kept in locked cages, then there is some measure of deterrence. If it is locked in a vault or solid building and access to the building is severely limited then the asset has more deterrence.

Finally, what is the perception of law enforcement in the area? Does the presence of law enforcement or guards act as a deterrent?

RISK RATING OR EVALUATION

When the assets are known and understood and their value is determined and after an assessment has been made of the aggressors, who they are and what their goals are, and finally, after the asset has been reviewed against the aggressors in light of the elements of visibility, usefulness, availability, accessibility, past and future potential incidents and deterrence, the risks have been assessed.

The facility now knows which assets have more risk than others. And the facility has begun to understand which threats are more likely. Then, the risk assessment is complete.

THE DYNAMIC NATURE OF RISK

As can be understood from the above discussion on risk assessment the risk picture can change rapidly. However once a risk assessment is prepared, it is far simpler to revise it than to reassess all the risks at one time. Depending upon the size of the facility and the number of assets and value, a risk assessment can be a simple or a complex process. Since the goal is to measure the risks with the intent of determining an adequate level of security, the risk assessment is a tool for determining where security may need improvement. Only the persons responsible can determine if there is enough security. The next step, determining what types of facilities need what types of security begins in the next chapter.

ADDITIONAL RESOURCES

Protective Counter Measures is a company that specializes in corporate risk management and has several tools for performing the initial assessments. Their web site is located at www.protectivecountermeasures.com/risk-management.htm.

Microsoft Office® has a free downloadable asset tracking database template. This template can be found on the internet at http://office.microsoft.com/templates and by clicking on "asset tracking" in the "templates" search window. The application requires Microsoft, Inc. Access 2000® database program.

How Much Does a Risk Assessment Cost?

Whether management decides to take resources from within the organization and perform the risk assessment or whether it can be contracted to consultants is something only management can address. Security consultants generally charge by the hour or week for preparation of a risk assessment and some organizations with more experience will be able to perform the tasks more quickly than others. Risk assessment teams can be small one or two person teams or larger; however, group dynamics suggest that a team larger than 7 will be ineffective. Usually teams have an odd number of members or a facilitator has the authority to break any ties in an analysis. Costs for risk assessments depend upon the sophistication of the tools used and the size and complexity of the assets. For a complex facility with multiple locations with a net worth of over $100 Million Dollars an initial risk assessment should not take more than 3 months and should not cost more than $200,000. For smaller, single unit facilities such as a commercial operation, the size of the risk assessment team is smaller and the amount of data and analysis is also correspondingly smaller. For a $5 Million Dollar commercial facility in a major city the risk assessment should not cost more than $40,000.

Chapter 3

Physical Separation: Fences, Barriers, Gates, Distance, Lighting

PHYSICAL SEPARATION: THE FIRST LINE OF DEFENSE

The first line of defense for security is physical separation of the assets from the threats. Most facilities accomplish this by separating the facility itself. This is done by selecting a site removed from the public. Factories, chemical plants, and power plants are often physically separated from the public by being in a remote location. There is a clear zone between the property boundary and the facility with a fence to mark the boundary. A fence is also used within an area to further enhance security of that area. Fences, walls and barricades keep the public out of the facility and control access of the people who visit the plant.

Other methods exist for physical separation within the complex which will be covered in Chapter 4 on buildings and building types. This chapter is primarily concerned with separation of the property from the public.

On a large complex, it can be uneconomical to fence the perimeter and in these cases the mere physical distance is a factor. The open space between the boundary and the assets is sometimes monitored by closed circuit television systems or the perimeter is sometimes protected by guards and roving patrols. This chapter discusses separation, distance and the use of different boundary markers like fences, walls and barricades. It also explains gates and openings through these boundary markers. Closed Circuit Television Systems are explained in Chapter 11 and Guards and Guard Forces are explained in Chapter 14.

For some facilities, it is not feasible to separate the facility from the public. Restaurants and commercial retail establishments need to be close to their customers and a large clear zone between the property boundary and the facility is not practical. Many commercial facilities have a small clear zone around the perimeter for fire fighting equipment. This clear zone is wide enough for a fire truck with enough room for another fire truck to pass by or 15 to 20 feet.

SEPARATION DISTANCES

As with any security the determination of separation distance depends upon the threats to the assets. In general, the greater the distance, the easier it is to maintain security. The risk assessment is an excellent tool for this determination. For a facility where the greatest threat is theft or vandalism less separation is required than for a facility where the greatest threat is a terrorist car or truck bomb. A facility with a boundary that physically separates itself with a distance greater than the "blast" zone will have adequately mitigated this threat. While it is recommended that for a "blast" threat an analysis be performed, the general guidance for stand-off distance has been recommended to be between 30 feet and 150 feet (10 to 45 meters) with the shorter distances being separation between less critical facilities and the longer distances being separation for more critical facilities.

Stand-off Zone

When the perimeter of the facility is delineated with a boundary like a fence, and the facility itself is a building or other structure, the distance between the two is called the stand-off zone. Depending upon the threat, this area can be maintained free of vehicles and people and the security force can patrol the area and prevent access unless it is properly authorized.

Visibility

For some threats, it is advantageous to protect the facility from line of sight. That is, the design of the fence or barrier is

For the specific terrorist threat of a car or truck bomb, the Federal Emergency Management Agency, or FEMA, has posted guidance on their web site. This manual is called the "Reference Manual to Mitigate Potential Terrorist Attacks against Buildings" and was published in December of 2003. The manual is located on the World Wide Web at http://www.fema.gov/pdf/fima/426/fema426.pdf. The manual is in Adobe® format with many graphics. The blast radius clear zone for stand-off distances is shown below.

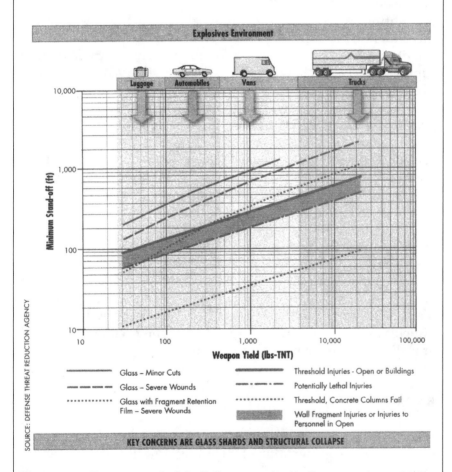

Figure 3-1. Recommended building stand-off distances for vehicles suspected of carrying bombs.

such that it blocks the view of outsiders, making visibility of the assets more difficult for those outside. The area of the clear zone itself should be free of things that could conceal an intruder like bushes and shrubbery. Walls or fences along the boundary should be managed to prevent the growth of weeds or piles of rubbish that could be used to conceal an intruder breaching the fence.

Lighting

To further protect the facility, perimeter lighting is used to illuminate any clear zones or stand-off zones. In general, the goal of lighting is to make the lights glare into the eyes of the intruders, while leaving the facility shaded to conceal the activities of the facility or guards. Lighting may also be needed to support the closed circuit television system. Chapter 11 discusses closed circuit television in more detail.

A facility may also want to use lighting to illuminate parking lots at night.

There are several ways to illuminate a facility at night and consideration should be given to the height of the light fixtures, the life of the fixtures and brilliance of the light fixtures. These factors, available from lighting vendors, determine the spacing and ultimately the number of light fixtures. Many vendors have software computer programs that will automatically determine spacing and height of parking lot and perimeter light fixtures based upon the brilliance and color of the light fixtures.

Lighting is also affected by the materials in the parking lot or the clear zone. If the materials are dark and light absorbing, like black asphalt or bare dark earth, intruders will be more difficult to see than if the background is lighter with pale colored gravels, concrete, or grass.

BARRIERS

In addition to fences and clear zones, barriers are a security measure that can be utilized to protect the facility. There are many types of barriers both natural and man-made. The type of barrier is chosen depending upon whether the threat is persons or vehicles.

The perimeter can be protected by natural or man-made barriers including bodies of water, natural depressions, hills or cliffs. However the facility does not want to be compromised because it is adjacent to higher terrain where assailants could take advantage of the overseeing position.

Man-made barriers include fences, walls, dikes, curbs, berms, depressions and trenches. These can be used to make it difficult for the intruder to cross the barrier. But it is important to recognize that not all barriers protect against all intruders. A boat or a person can cross a body of water, while a car or truck cannot.

One common type of portable barrier is the Jersey Barrier. The Jersey Barrier is composed of large concrete blocks securely anchored to the ground and linked together. The Jersey Barrier is often seen adjacent to highway work. A photograph of a typical Jersey Barrier is shown in Figure 3-2.

Barriers, like the clear zone are designed to prevent the threat. As a result some barriers are shaped differently from others. One method of preventing access by vehicles while allowing people access is by installing devices that are called bollards. A bollard is a post set into the ground such that a

Figure 3-2. A jersey barrier is a good temporary security measure to control traffic flow. (*Photo by Robert Reid, 2004*)

vehicle cannot pass. Bollards, also sometimes called pipe bollards, are seen at large stadiums, and in crosswalks.

A facility can also use a curb to prevent vehicle access. In order to prevent vehicles from crossing, the curb should be at least 8 inches high. Obviously a curb will not prevent access by pedestrians.

Finally, a method for preventing vehicles from crossing the barrier can be a large cable stretched between posts. To prevent a vehicle from penetrating, the cable should be at least 3/4" thick and anchored in the ground at the end points.

Combined Barricades

A facility that wants to keep out both pedestrians and vehicles can combine the types of barricades. For example a chain link fence could have a vehicle resisting cable woven into the mesh, a fence or wall could be protected by a berm, dike or curb. Or a Jersey Barrier could be erected next to the fence.

Figure 3-3. Pipe Bollards protect entrances to facilities by vehicles. (*Photo by Robert Reid, 2004*)

FENCE

There are many types of fences for protection of property. A fence is a statement by the facility that the boundary is defined. However, a fence, by itself, is not a very effective method to prevent intruders. The U.S. Army manual on physical protection indicates that a 7-foot fence (without wire overhangs) can be climbed in 4 seconds, and it can be cut in 10 seconds. So a chain link fence is more of a boundary marker than a deterrent.

From the previous discussion of risk assessment in Chapter 2, the facility manager knows that there are two different situations involved in protecting an asset. In the first case the intruder wants to enter the facility and steal the asset, while in the second case, the intruder's goal is to enter and destroy the asset without trying to escape. The security plan should take these alternates into account because some facilities find it favorable to have the fence delay the intruder's escape while in the other case the fence/barricade is needed to prevent intrusion by even the most dedicated intruder.

Solid fences, while blocking visibility, also should be designed to resist wind forces. A solid fence can be blown down in a strong windstorm. To prevent this problem the posts must be set closer together and embedded deeper. For most fences posts are placed on 10- to 15-foot centers and are embedded in concrete. The general rule of thumb for fence posts is that the diameter of the hole is at least three times the diameter of the post. Unless the ground under the fence is very hard, for security purposes, it is common to install a bottom rail, or a curb under the fence. For a high security installation the wire mesh is extended down into the ground to prevent tunneling.

Barbed Wire

The most common fence type is barbed wire and while three or four strands of barbed wire will contain animals, it is not capable of slowing down individuals or even vehicles very much. A 7 or 8 foot barbed wire fence with a horizontal strand every 6 inches, posts every 8 feet and a vertical strand halfway between the posts is more effective. Barbed wire fences need to be constructed in straight lines with heavy reinforced corners. A curved barbed wire fence puts tension from the wire onto the posts and can cause the posts to fail prematurely.

Chain Link

The chain link fence (Figure 3.4) is more effective than barbed wire and is a very common barrier. It is economical with a unit cost of approximately $8.00 per linear foot, installed. Chain link fence is woven of zinc coated wire mesh, usually with 2 inch by 2 inch holes. The zinc coating prevents the wire mesh from rusting. There are two common gauges for the wire mesh, 9 gauge and 11 gauge with the 9 gauge being heavier wire. Generally, the 9 gauge is recommended for industrial/ commercial applications while the 11 gauge is recommended only for commercial applications. The heavier size of the mesh does not indicate more security but affects the life of the fence. Depending upon weather conditions the life of a chain link fence is 25 to 30 years. The posts are usually zinc coated as well and are usually 10 to 20 feet apart. The fabric is held in place with special fasteners fabricated for the wire mesh fabric. At the top of the fence a bracket is installed which is designed to hold barbed wire and overhang the fence slightly to make it more difficult to climb. These overhanging brackets are called barb arms or outriggers and they are slotted for holding barbed wire. A metal pin is inserted into the outrigger to hold the barbed wire in place.

Chain link fabric has to be tensioned to place it on the fence. And the fence can have top rail, bottom rail and mid rail or the top and bottom can be tensioned with a cable. At corners and gates heavier posts are required to hold the tension. Chain link fences using cable at the top and bottom are more difficult to climb than chain link fences with top and bottom rails.

Variations on chain link fence include coating the fabric with vinyl or plastic to make it more appealing and for privacy; wood or plastic slats can be inserted through the wire mesh to obstruct visibility.

Barbed concertina wire can also be placed atop the fence for added security. This wire comes in coils and it is specially designed to "hook" intruders. The barbs are cut out of flat tape and can be placed on top of chain link fence or stacked on the

Figure 3-4. Chain link fence is an effective barrier for marking a boundary. This figure shows a typical 5-foot-high chain link fence being used for a commercial storage facility. (*Photo by Robert Reid, 2004*)

ground. The ability of the wire to restrain an intruder is a function of the diameter of the wire coils with a large diameter providing greater resistance. Figure 3-5 shows Razor Wire® a barbed tape wire for increased security. Finally, some local codes prohibit the use of barbed or concertina wire at the top of a fence because it is a safety hazard.

WireWall™

A new welded wire fence called WireWall™ is manufactured for facilities that have a higher security application than can be met with chain link fence. The fence is made from welded wire in either 8 or 10 gauge with smaller openings than the 2-inch mesh of chain link fencing. The wire is welded at intersections instead of woven. The WireWall™ is more expensive but it is also more difficult to climb and to cut than chain link. The manufacturer reports that while a chain link fence can be cut in a few seconds, it takes a few minutes to cut through a WireWall™

Figure 3-5. Razor Wire®. Reprinted with permission: Razor Wire International Copyright 2004 Phoenix, Arizona. All rights reserved.

For very high security installations a perimeter fence is two 7 or 8 foot high chain link fences about 20 feet apart, each with three strands of barbed wire overhanging both the inner and outer fence. Sometimes concertina wire or razor wire is used on top of the fence in addition to the barbed wire. The inner zone between the fences is called the kill zone, it is the zone where the intruders will be forced to surrender or, if not, they can be shot. The area between the fences has sensors to signal if anyone is climbing or tampering with the fence and other intrusion detection systems are provided to alert an armed around the clock emergency response force. The wire mesh of these high security fences is buried up to 4 feet into the ground to prevent intruders from tunneling under the fence, and steel cable is enmeshed in the fence to prevent it from being knocked down by a vehicle traveling at high speed.

fence. The material comes galvanized or plastic coated in a variety of colors. Because WireWall® is proprietary it is only available from one manufacturer in Northbridge, MA. Figure 3-6 shows some details of the WireWall® fence.

Wrought Iron Fence

In addition to the chain link fences there are fences made from wrought iron. These fences are used more commonly in public areas because they are more difficult to climb than chain link fences and they are more aesthetically pleasing than chain link fencing. However, the spacing on wrought iron bars is usually greater. These fences are generally seen at ball parks, stadiums and public swimming pools. They are more difficult to climb than chain link fences but they are also more expensive.

Most modern "wrought iron" or "ornamental steel" fence is constructed of tubular steel or aluminum members in sizes from 20 mm (3/4 in.) square to 100 mm (4 in.) square. The fence is fabricated in a shop and the sections are brought to the field and tacked or bolted between posts in place. These types of fences are built with metal or masonry posts on 8- to 12-foot centers with two or three horizontal rails.

For swimming pools and other water attractions, the wrought iron fence is tested using a 6 inch ball pressed between the bars and at the base. The test is to protect small children from sticking their head between the bars and becoming stuck. These fences can be painted, coated or galvanized and they will last longer than chain link or welded wire mesh fences.

Depending upon how the fence is constructed, it is possible that a wrought iron fence can stop a small vehicle. If the fence has a curb 8 to 12 inches high underneath, it can be a very effective barrier against a moving vehicle. However, this type of fence does not prevent line of sight observation.

Wooden or Composite Material Fences

In addition to fences made from chain link, welded wire and wrought iron a facility can construct fences of wood or materials fabricated to look like wood. Wood fences are more commonly associated with residential construction but wood has been an effective barrier. A wood fence, properly constructed, obstructs

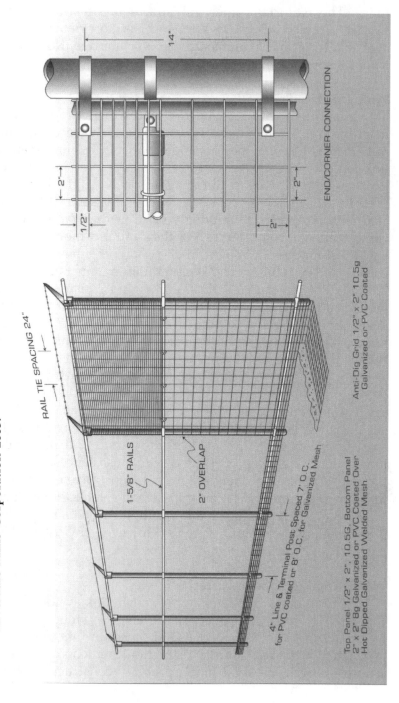

Figure 3-6. WireWall™ Fence. Reprinted from Riverdale Catalog with permission. Copyright Riverdale Mills Corporation 2003.

Figure 3-7. Typical wrought iron fence. *(Photo by Robert Reid, 2004)*

the view of people outside the fenced area. Wood fences are more expensive than chain link fences, and they require more maintenance. In addition the wood material common for most fencing applications is not very strong and it can be kicked in or broken though relatively easily. There are a number of new hybrid materials that emulate wood fencing that are stronger and last longer. These are also more expensive than regular wood fencing but are more economical in the long run because the fence lasts longer and requires less maintenance. These materials are fiberglass or plastic materials, often made from recycled products. They are very tough and durable.

Unless the wood fence is protected by a dike or berm, it will not stop a moving vehicle that attempts to crash through it.

WALLS

Perimeter protection can be accomplished with walls instead of fence. While walls are much more expensive than fences, they require less maintenance, last longer and provide better security

Figure 3-8. A typical wood fence. These fences do not provide a great deal of security, but they do obstruct vision, preventing intruders from knowing location of material or equipment inside the area. (*Photo by Robert Reid, 2004*)

than fences. Walls can be constructed of all building materials but most commonly walls are constructed of masonry block or concrete. Walls are an effective barrier to the pubic and some walls are 20 feet tall. They are difficult to climb, and if reinforced they are difficult to breach. A well built masonry wall will stop a vehicle. Because the wall is much heavier than fence it requires additional sub grade work to support it, and usually must be constructed to below the frost line to prevent frost heave from buckling the fence. Some poorly constructed walls settle and crack but they remain effective as a security barrier as long as they do not collapse. The construction of a masonry or concrete wall requires some structural analysis. It has been the practice to reinforce walls with steel where they are adjacent to pedestrian walkways, as they are prone to complete collapse in an earthquake.

The top of the masonry or block wall usually has a cap stone placed on it to drain water and prevent water from penetrating

Figure 3-9. Block walls provide good perimeter security, but they are expensive. (*Photo by Robert Reid, 2004*)

the brick. This adds life to the wall. In some third world countries, the capstone has shards of glass embedded in it to discourage climbing over the wall; however, this practice would probably subject the facility to some liability since it would injure a person climbing the fence.

GATES

Fences, walls and other boundaries require gates to allow the passage of vehicles and personnel. Gates can be the most frustrating part of a fence or wall because they will require maintenance and when they are not working properly, they will be left open as the process of constantly opening and closing the gate becomes too difficult for the staff. Gates for personnel are different from gates for vehicles and consideration should be given for the number of people/pedestrians using the pedestrian gate and the number and size of vehicles using the vehicle gates.

Vehicle Gates

Vehicle gates can have a power mechanism for opening and closing, they can be tended by guards who open and close the gate manually, or they can be opened and closed by the vehicle operator or a passenger. There are many mechanisms for opening and closing gates from hydraulic operators to electric motors. Electronic control of gates will be discussed in Chapter 8 and power supply for gate operators is explained in Chapter 13.

Gate material can be the same as fence material, for example, the chain link fence can have a chain link gate, and a wrought iron fence can have a wrought iron gate. In general, for vehicles, gates are either sliding or hinged.

Sliding Gate

The sliding gate usually runs on a small trolley of wheels at the bottom or top although for most gates wheels on the bottom are preferred. The wheels hold the gate up and allow the gate to slide back along the fence, creating an opening for the vehicle. For security, a partially open gate is vulnerable and even the most elaborate crash gate can be penetrated if struck when it is opening or closing.

If the sliding gate is powered it will usually have a chain operator that pulls the gate open and the chain motors are reversed to pull the gate closed again. Non-powered gates are simply pushed open and then pulled shut again. Sliding gates with bottom mounted rollers often have problems with the track because the track becomes full of dirt, dust, debris, rocks or sticks. This lack of maintenance will cause the gate to stick making opening and closing it more difficult. For this reason some gates have an overhead track. For gates with an overhead track, the gate hangs from the wheels allowing it to slide open and closed. The problem with the gate that has an overhead track is that the vehicle that has a load higher than the overhead track cannot go through the gate. Shipments to the facility have to be coordinated to make sure they are not loaded to a height greater than allowed by the gate. In some facilities, there is a truck pull out next to the gate that allows vehicles to be off loaded outside the gate and the material brought into the facility with the facility's forces.

Figure 3-10. A sliding gate is an efficient barricade. This photograph shows the gate chain operator for a sliding wrought iron gate. Note that jersey barriers have been placed to force a vehicle to approach the gate directly. (*Photo by Robert Reid, 2004*)

Hinged Gates

The problems with sliding gates are eliminated with hinged gates. A hinged gate opens either into the facility or outward and can be both double, that is, opening in the middle and hinged at each end or the entire gate can be hinged at either end. Opening a hinged single gate is easier than a double gate and takes less time, but a single wide gate, sometimes up to 20 feet, can put large forces on the hinges and they will require more maintenance over time. Some hinged gates have rollers and a track on the end, reducing forces on the hinge. Figure 3-11 shows a double hinged gate. More room is necessary inside the facility for opening the hinged gate than the sliding gate.

Crash Barrier

Because gates are made from fence material, it is not feasible to make them strong enough to act as a crash barrier. As a result a crash barrier gate is sometimes installed in addition to the gate. A crash barrier consists of one or more large structural beams that are seated in heavy concrete pockets at each end. Most crash

Figure 3-11. A typical chain link double hinged gate. (*Photo by Robert Reid, 2004*)

Figure 3-12. Crash gate for a police compound. (*Photo by Robert Reid, 2004*)

gates have a counter weight on one end that makes opening and closing the gate much easier. These gates can be manually operated by the guard force or the vehicle operators or they can be powered with hydraulic or electric motors. The design of the crash gate should be based upon the results of the threat assessment, i.e. the heavier the gate, for the more serious threat. Figure 3-12 shows a fairly light crash gate.

In general a crash gate does not prevent pedestrian traffic. Its sole purpose is to stop a large moving vehicle.

PEDESTRIAN GATES

While a vehicle gate can be used for pedestrians, the risk assessment assumes that most people who do not need to drive into the facility should park outside and walk in. In this case, traffic through the vehicle gate by opening and closing it for one person after another is hard on the gate operator. Usually a gate designed specifically for individuals is better suited for the task. These gates can be powered or manually operated and, depending upon the threat, the gate can be tended by guard or locks or keys provided to staff trusted to open/close and lock their own gate. Many pedestrian gates are hinged and work with a lock and key. Chapter 6 discusses locks and access hardware for doors and hardware that can be applied to doors can be applied to gates as well.

Figure 3-13 shows a typical chain link hinged gate.

Other Types of Personnel Gates

Several other types of personnel gates are used for higher or lower security applications. A typical turnstile gate is used to admit people to large facilities and these gates also act as personnel counters to keep track of the number of tickets sold against the number of persons in the facility. Turnstiles are used in subways, train stations, and at stadiums and theaters.

A more heavy duty gate is used for high security applications like detention centers. The turnstile gate shown in Figure 3-14 shows a high security turnstile. These gates have settings at three positions and can lock at the mid point, holding

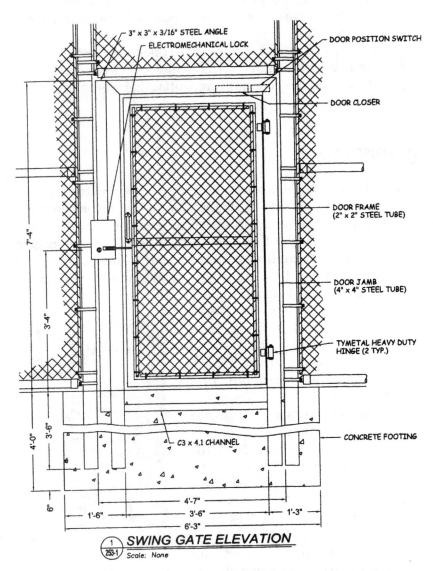

Figure 3-13. Personal chain link hinged gate. Reprinted with permission; copyright Tymetal Corp., all rights reserved; Greenwich, NY.

the person until additional security arrives. These gates can be installed inside or outside but these require a lot of maintenance to operate properly.

Both types of turnstiles can be set to operate to restrict access in or out or in both directions. Usually these gates serve to detain the person passing the gate until credentials have been identified and confirmed.

GUARDHOUSES

Finally, in many instances securing the perimeter requires the use of guards. Guards check the identity of persons entering or leaving the facility and can search selected staff, all staff and the vehicles. Where the threat is deemed necessary to provide guards at the perimeter, a guardhouse is necessary. The guardhouse provides heat and light for the guards, a place for perimeter patrols to stop and check in, and many other functions including a toilet. Depending upon whether the decision is to have armed guards or just parking lot attendants, the design of the guard house is affected accordingly.

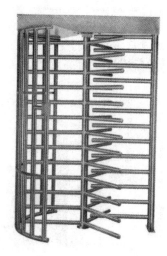

Figure 3-15 shows a guard house for an apartment complex.

Lighting must be provided for the guards to see the necessary identification, and power and communications should also be provided. Many guard houses have their own power supply and on high security guard houses, even the ventilation intakes and exhausts are protected by blast plates.

Figure 3-14. Personal turnstile gate. Reprinted with permission from Alvarado Manufacturing Company, Inc., Chino, California.

Figure 3-15. Guard house for an apartment complex. (*Photo by Robert Reid, 2004*)

Form: Things to know about your guard booth.

1. How will the guard booth be used by the guard staff? Are restrooms and eating planned?
2. Does the booth check vehicle traffic as well as pedestrian traffic? If the gate is for vehicles: is there an overhang that can be struck by vehicles? Does the overhang protect the guard from weather?
3. What is the threat level to the guards in the guard booth? Is there danger that the guards could be attacked? If the guard house is the first line of defense is it protected with bullet proof glass and walls?
4. As owner, do you care what the guard booth says about your facility? Does it need to be architecturally compatible?
5. Is the booth for vehicles or walk in traffic? Are the people and vehicles subject to search? Will all vehicles be searched or just vehicles without identifying stickers?
6. Should the booth be a prefabricated unit purchased from a commercial vendor or does it need to be traditional construction? Note: there are different tax implications for portable booths over permanent ones.

CONCLUSION

The perimeter is the first line of defense. Clear zones, fences and walls are all elements of a good security program. Fences delineate a boundary more than prevent access. A determined intruder can penetrate the fences eventually. But these perimeter defenses can delay an intruder and provide the first step to advance notice that the threat level has increased.

The next layer in the security picture is the facility itself—the buildings. The guide for securing buildings begins in the next chapter.

RESOURCES

Reference Manual to Prevent Terrorist Attacks against Buildings: Manual 426, Federal Emergency Management Agency December 2003. http://www.fema.gov/pdf/fima/426/fema426.pdf

Robert N. Reid, Park Fencing, *Construction Specifier Magazine* Vol. 46 No. 12 December 1993.

Physical Security FM3-19.30, United States Army January 2001.

Boundary Fence and Railing Systems, Inc. www.boundary-fences.com

WireWall™ Fence Systems www.riverdale.com

Tymetal Corporation, Security Gates www.tymetal.com/index.htm

Alvarado Manufacturing Company, Inc. www.alvaradomfg.com

Chapter 4

Building Elements and Explosion Behavior

With a complete risk assessment and knowledge of the functions within the building and with the perimeter secure, the next step in securing the facility is to secure the enclosure. In order to accomplish this element a basic understanding of the building's construction is required. Later chapters address entrances and openings and controlling them. This chapter explains the construction materials of the building being secured and why one type is favored over another from a security standpoint. This chapter examines building frame elements, cladding and roofing and examines how these elements behave in an explosion.

TYPES OF CONSTRUCTION

Most buildings are designed based upon what the people will be doing inside. Therefore, one of the main drivers for the type of materials to be used in construction is the building fire code. For various types of buildings, the fire code determines certain fire resistive elements required for the building's construction materials. These elements also tend to improve the building's security. A building can be constructed from thin sheet metal over a frame, as in a warehouse, or it can be a masonry building constructed using brick and mortar as would be used for a school.

While the fire code determines some materials of construction, heating and cooling of the facility affects the building's materials of construction as well. A thicker/more insulated cladding and roof will reduce energy losses. These elements also assist in helping to make the facility more secure.

Most buildings consist of a frame either of steel, concrete or wood which is then covered with a cladding. Cladding is a lighter material that fills in between the frame pieces and provides protection from the elements. The frame, cladding, roof and floors make up the enclosure.

FRAME

Most modern commercial and institutional buildings consist of a metal frame with one or more types of cladding. (Cladding is the skin on the outside of a building.) The frame can be made from structural steel, brick or block masonry, reinforced concrete or wood. Some unique new types of construction materials have been tried for residences but are not in general application elsewhere and are not addressed here. For security analysis, one type of construction material can be subject to more threats than another type.

Steel Frame

The most economical building in terms of construction materials is a steel frame with a cladding system. Depending upon the designer, the cladding can be metal, brick, glass or a combination of all three. In modern buildings, the cladding is not holding up the structure, the metal frame performs that function. The steel frame can be bolted or welded together. Provided the columns are not cut at or near the base of the structure, this type of frame is very robust from a security standpoint. The frame is relatively blast resistant and the building elements will not burn; however, they can be severely weakened by fire. As a result most structural elements in a steel frame are treated with a fire retardant material to protect the metal from extreme fire temperatures. For buildings taller than about 4 stories, the steel frame is the most economical choice.

Brick and Masonry

Older buildings used masonry or stone as the main structural element for the walls—then beams were laid across the walls to make the roof. Brick Masonry is excellent security material—it is difficult to penetrate. It is difficult to bore holes into these types of walls. And if the brick and masonry has been reinforced with steel, these structures will have good blast resistance characteristics. Unreinforced masonry is not as blast resistant and could completely crumble if shaken by an earthquake or large explosion. Adobe brick is another example of a masonry material. In earthquake zones, large buildings of adobe have had to be reinforced with steel rods or bars to strengthen them so they will not collapse in an earthquake.

Reinforced Concrete

From a pure security standpoint, reinforced concrete is the most secure of the buildings. The concrete won't burn, it is thick and difficult to penetrate, and it is blast resistant. Yet reinforced concrete is expensive and somewhat height limited. The cladding can be concrete, but it requires more massive framing which increases cost.

Wood Frame

In general, commercial and institutional buildings are not made from wood, although some office and medical facilities are framed from wood. These buildings have low occupancy and are easy to evacuate in the event of fire. Because wood is combustible, it can be treated to make it fire resistive, but this treatment is expensive causing designers to opt for the metal frame instead. Wood is not as strong as the other construction materials. In general it can be more expensive than the other construction materials.

Other Types of Frame Materials

There are other types of construction materials although not commonly used in the frame. One has been the use of Styrofoam block for the walls of residences. This material has not been on the market a long time, and the long-term performance of this type of material has not generally accepted. Although Styrofoam can be treated so that it will not burn, it can give off vapors in a fire that can be hazardous to health.

Some homes have been constructed of reinforced earth. In this method of construction, soil that has been treated to prevent the growth of organisms is hammered into blocks that are erected as the walls. Not many large buildings have been constructed of this material but it will not burn. The blast resistance of this type of structure is unknown.

Combinations

All of the materials can be combined to form a structure, wood combined with masonry, steel and reinforced concrete, wood and steel, etc., all with varying degrees of success. For security the more fragile the frame is and the more combustible the frame is, the greater the risk in the security analysis.

CLADDING

Cladding is the material that forms the skin or shell of the building. The doors, windows, and other openings are placed into the material. From a security standpoint, it is important to know and understand the facility cladding system because some are weaker than others. This means that the cladding system may not be as strong as suspected and it can be easily penetrated. The size and shape of windows and doors also plays a role in the building design. From a modern perspective, the current technology of heating, cooling and air handling equipment allows buildings to be constructed without windows. There may be windows for aesthetic purposes, but modern air conditioning design favors a building without windows because windows allow sunlight and makes the rooms warm up and cool down in different areas of the building at different rates. This would cause occupants to want to prop doors open for example, which is why this could lead to security problems. However, a windowless building relies upon artificial lighting and some people prefer some natural light instead. Employee health and productivity are also a function of natural light. Hence, designers provide windows.

Openings into buildings are one of the weaker elements and the next chapter discusses openings and the protection of them in detail. Some building cladding systems are discussed below with particular attention paid to their effectiveness from a security standpoint.

Metal Panels

One cladding material is composed of thin sheet metal panels. This material is typical of metal warehouses. The metal panels are selected according to thickness with the higher gauge numbers thinner and more fragile than the larger numbers. Typical metal frame panels vary from 22 gauge to 10 gauge (0.0336 inches to 0.1382 inches). These metal panels will not burn, but they are easy to penetrate.

Masonry/Block/Stone

Just as in the frame, the cladding of a building can be made of masonry or block. It has good insulating characteristics, it is difficult to penetrate and it will not burn. It will, however, fracture unless it is reinforced with concrete or steel.

Some facilities are cladded with thin sheets of cut stone like marble or granite. The thinner material is better from the viewpoint of the building constructor because of there is less weight. Usually these materials are adhered to a frame or other wall assembly before being placed. It makes the facility appear to be constructed of large stone blocks, but it is not. Security of this type of cladding is a function of the thickness of the material and how it is attached to the frame.

Wood

Wood has been used as cladding for many buildings. The wood can be treated or painted to make it fire retardant. However, wood will burn and depending upon the paint the paint may also burn. Depending upon thickness, these walls are difficult to penetrate.

Concrete

Concrete is an excellent cladding material and many one story buildings are constructed of tilt up concrete panels. Concrete will not burn, but it is heavy and it must be reinforced as a cladding to prevent cracking.

Glass

Many beautiful buildings are constructed of a steel frame with a complex glass panel assembly cladding. Glass is easy to penetrate, shatters but will not burn. It is important to understand what type of glass the facility has for the security analysis because different types of glass behave differently under the same situations. For example, the glass in automobile windshields is far different from the typical glass window in a house.

Studies report that over 85% of injuries in a bomb blast are from flying glass. The bomb causes a blast wave that shatters and implodes the glass, sending flying fragments throughout the room. To prevent flying glass, windows can be treated with film that holds the glass and keeps it from flying into the building. In addition, installed draperies will capture glass shards and prevent them from flying into the rooms. Float glass can be removed and replaced with tempered safety glass. In the U.S. Army's technical manual to prevent terrorist attack, window glass can be removed and replaced with plywood. This would obviously be a temporary measure, but in an escalated threat situation, if time permits, this would be a good temporary fix.

Heat-tempered glass will not shatter into razor sharp slivers as annealed glass does. Instead, it shatters into small cubes. However, while this type of glass is better from a security standpoint than annealed glass, the small fragments can become high velocity projectiles in a blast.

Laminated glass is what is present in automobile windshields. It consists of two sheets of glass layered onto a thin polycarbonate material that holds the glass in place if penetrated or exposed to a blast. It is much more effective than annealed glass but it is also more expensive.

In addition to glass, clear or shaded polycarbonate materials can be used instead. Bulletproof glass is, in fact, a polycarbonate material or a combination of glass and polycarbonate.

Window design is more complex than just the material in the frame. More information is provided on window design and security in the next chapter.

Metal Sandwich Panels

A new type of cladding for some structures is the prefabricated insulated metal panel. This type of panel consists of two thin metal sheets separated by polyisocyanurate foam. These panels are more rigid than the metal panels themselves and they are fire retardant so they will not burn. They are easily penetrated but more difficult to penetrate than wood. While generally they are not used as frame support they are strong enough to support roof beams. The method of installing a window into a wall fabricated of these metal panels is slightly complex and more complex than installing a window in a wood cladding.

Exterior Insulated Finish System

For cladding of buildings there is an assembly called an exterior insulated finish system (EIFS) that is a composite system that looks like stucco on the outside and is sometimes referred to as "synthetic stucco." EIFS is an insulating board over building sheathing, followed by a protective weather barrier. It may appear to be stucco but it is not. Depending upon the insulation board and the waterproofing membrane material and adhesives, this material will burn although it is treated to be fire retardant. It is easily penetrated, but because the insulation board overlays the sheathing, it is stronger than some of the other cladding materials.

ROOFING

As with cladding, the roofing system is a waterproof or water shedding membrane that is laid over a structural frame on the top of the building. The roofing system lays on a structure that is designed to hold it. For example, a roof can be made from trusses with a wood sheathing on top of the trusses and then layers of asphalt paper and shingles on top. Roofs are divided into two categories, the low slope roof which is nearly flat, and the high slope roof with a steep pitch. The underlayment is called the deck. The water shedding or water proofing membranes lay on the deck. Low slope roofs can be made from wood, metal or concrete. Obviously the more fire resistive the deck, the better it is. A low sloped roof often has a parapet wall (Figure 4-1). The parapet acts as a wind break for the roof to protect it. For security reasons, a sloped or curved roof is favored over a flat roof with parapet walls. A sloped roof has a tendency to deflect a shock wave blast upward and it has a tendency to shed incendiary devices.

Figure 4-1. Parapet wall. The problem with this type of wall is that the wall can catch explosive devices that will fall off of a sloped roof. (*Photo by Robert Reid, 2004*)

In addition to the roof shape, some types of roofing materials are better than others are from a security standpoint. Materials that will not burn perform better than those that will. Most roofing systems have a coating of inert material like stone to aid in weathering. Roofing also is required to have flame spread and fire retardant characteristics to prevent the spread of fire and prevent fire from jumping from rooftop to rooftop. However, roofs of clay tile and slate have more fire resistive characteristics than asphalt shingles. Some low slope roofing materials are flame retardant, while others, like bitumen (also called roofing tar), burn more readily.

Shingles

Asphalt shingles are difficult to penetrate and resistant to fire. However, the organic compounds in the shingles will eventually burn. There are other shingle materials like fiberglass, clay or slate shingles that can be more or resistant, depending upon the materials.

Tar

Modern roofing materials in the USA use a material called bitumen or modified bitumen to provide the waterproofing layer. These are usually covered with gravel to provide weathering and fire resistive characteristics. These materials, although treated to be fire resistive, are combustible.

Rolled Roofing

Rolled roofing is very similar to asphalt shingles; however, instead of being applied as shingles, it is rolled onto the roof. Again, it is combustible.

Elastic or Plastic Materials

There are a number of roofing systems that consist of a plastic or rubber like mat that overlays the support system. On some of these mats, the insulation is above the waterproofing layer and on others; it is below the waterproofing layer. These materials will burn, but there is less combustible material here than in the rolled roofing. Some of these mats are protected with loose stone ballast. In a catastrophic event, the stones can fall through the roof onto rescue workers.

Other roofing materials are adhered to the deck or insulation boards with adhesive.

Metal

Finally, metal panels are used on roofs in the same way they are used for cladding. Usually the metal roof panels are lighter than cladding panels. The metal will not burn but a sharp object can penetrate it. Depending upon the slope of the roof, metal panels may provide some protection by shedding incendiary devices.

LOCATION

A building can be above ground, multi-story or single-story and it can be wholly or partially buried. For top security the underground building is obviously best. It cannot burn and its walls are protected with earth, but it is expensive and the lack of windows does not appeal to many staff.

Solar energy design applications have called for buildings to be partially buried, usually on the north side (in the northern hemisphere) leaving the south side exposed to sunlight. This building has advantages that its walls cannot be penetrated from the North side. Often, however, the South side contains quite a bit of glass and glass can be a big problem from a security standpoint.

As discussed in Chapter 3, a location with a lot of clear space around it allowing for clear zones and stand-offs is preferred, however, some facilities do not have this luxury. Some buildings are downtown in a densely packed zone. For these locations, the construction is usually fire resistive and fire codes call for firewalls between buildings. These firewalls will prevent the spread of fire and help to improve security by limiting access between buildings. Large city complexes will usually have underground access.

A facility may face a risk if it is adjacent to a much taller building or across the street from a taller building where the other facility looks down on the first one. This gives line of sight visibility from the taller building onto the shorter building. Depending upon what assets need to be protected, line of sight may need to be limited. Billboards and signs can be erected that will limit line of sight.

The building assessment should consider trees and towers that are adjacent, again allowing line of sight. Some projectile weapons can be used to threaten a building security if line of sight is possible. In addition, trees and towers can be attacked and these fall onto the building.

EXPLOSIONS

Finally, in a blast from a car or truck bomb, the whole building could collapse. Collapse of a building is a function of the magnitude of the blast and can usually only be mitigated by increasing the stand-off distance. Some elements of the building can be reinforced but the blast pressure from an explosion can be as high as 10 pounds per square inch.

Shock Wave

In an explosion, the detonation sends out a blast wave or shock wave at the speed of sound. As the blast propagates, its force is diminished rapidly as a function of the distance. The pressures from a blast vary with distance and the size of the explosion. See Figure 4-2, which shows the typical forces from an explosion at various distances.

A complete analysis of blast pressures from explosions is relatively complex and uses computer programs to measure pressures and flows. If a facility warrants this type of analysis, a structural engineering firm may be hired to analyze blast zone effects. However, these analyses are expensive, costing thousands of dollars and the results are going to be somewhat imprecise since the analysis requires many assumptions.

Since blast pressures are what damages a building, Table 4-1 provides information on the result of some damages resulting from these pressures.

An analysis of blast pressures from buildings is complex and expensive. The analysis may be warranted if the building assets indicate such an analysis is required. The Department of Homeland Security is advising cities and towns of the relative risk of a terrorist attack with a car or truck bomb. Some funding may be available through local government as a result of Homeland Security support to US communities. For buildings in other areas, an analysis may be necessary, but may not be supported by local government.

MIXED AREAS

Within a building it is possible to subdivide the structure to provide enhanced security protection. The previous discussions

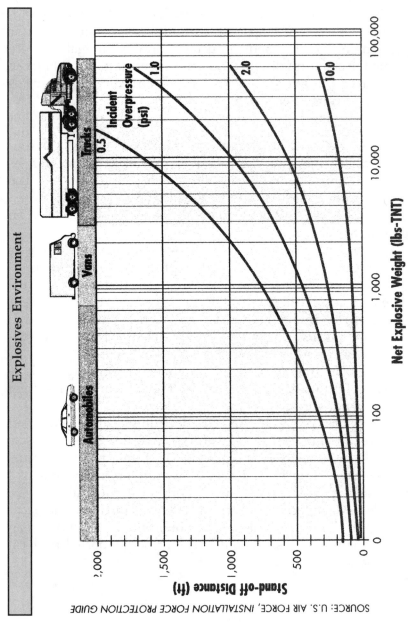

Figure 4-2. Incidents of overpressure measured in pounds per square inch as a function of stand-off distance.

Table 4-1: Resistance of building elements to overpressure from explosive blast.

Damage	Incident of Overpressure (psi)
Typical window glass breakage	0.15-0.22
Minor damage to some buildings	0.5- 1.1
Panels of sheet metal buckled	1.1 - 1.8
Failure of concrete block walls	1.8 - 2.9
Collapse of wood framed buildings	Over 5.0
Serious damage to steel framed buildings	4 - 7
Severe damage to reinforced concrete structures	6 - 9
Probably total destruction of most buildings	10 - 12

Source: Explosive shock in Air, Kinney and Graham, 1985; Facility Damage and Personnel Injury from Explosive Blast, Montgomery and Ward, 1993 and The Effects of Nuclear weapons, 3rd Edition, Glasstone and Dolan, 1977.

on cladding and blast effects indicate there is more risk to the rooms on the perimeter of the building than interior rooms, except in the case of total building collapse, whereby the risk is the same for the entire building.

However, within a building there are areas that can be more protective than others. This can be accomplished by strengthening the walls, floor, ceiling, and by eliminating glass. Objects in the ceiling; like light fixtures and heating, ventilating and air conditioning duct; can be reinforced.

Safe Area

A requirement of the Americans with Disabilities Act calls for an area in a multi-story building to be designated an "area of rescue assistance." This is an area near the exit stairway with expanded space for disabled persons to wait for rescue in the event a building needs to be rapidly evacuated. It also requires for signage, and two-way communication with the ground. Figure 4-3 shows general requirements for an Area of Rescue Assistance.

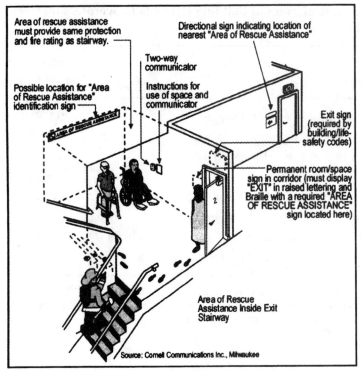

Area of rescue assistance must provide same protection and fire rating as stairway.

Directional sign indicating location of nearest "Area of Rescue Assistance"

Two-way communicator

Possible location for "Area of Rescue Assistance" identification sign

Instructions for use of space and communicator

Exit sign (required by building/life-safety codes)

Permanent room/space sign in corridor (must display "EXIT" in raised lettering and Braille with a required "AREA OF RESCUE ASSISTANCE" sign located here)

Area of Rescue Assistance Inside Exit Stairway

Source: Cornell Communications Inc., Milwaukee

A diagram of an escape route, which follows Section 4.311.4 of the ADA Accessibility Guidelines.

Figure 4-3. Area of safe rescue assistance. Reprinted with permission: *Security Dealer Magazine*, September 2003 issue. Copyright 2003 Cygnus Business Media, Jensen Beach, FL.

Safe Room

One area of the facility, on each floor or near the executive offices should be designated a "safe room." The safe room is designed to delay and deter the activities of an intruder or intruders. The safe room should have heavy doors, emergency lighting, and secure communications equipment. The intent is to delay the intruders until help can arrive. Secure communications equipment can be a cellular telephone or telephones with independent circuits. The doorway should be solid wood or metal set in a robust frame and lockable from the inside with a deadbolt lock.

For extra security slide bolts can be provided at the top and bottom of the door. The room should also contain a fire extinguisher.

Vault

Like the safe room, some facilities that have a need for high security may decide to have a vault. Although a vault is inside the building, it has its own high security door and concrete reinforced or steel walls. Important assets are kept in the vault. The vault should have its own lighting and ventilation systems, and an independent communications system.

CONCLUSION

Now that the elements of buildings are known, the next step in securing the facility comes from understanding the openings in the building that are the normal pathway for an intruder to try to access the facility. Openings include doors, windows and ventilation openings for air conditioning. Some buildings have special utility openings called a "chase." These topics are addressed in the next chapter, Openings in Buildings.

RESOURCES

Reid, Robert N. *The Roofing and Cladding Systems Handbook,* Copyright 2000 by Fairmont Press, GA. www.fairmontpress.com

Chapter 5

Openings in Buildings

INTRODUCTION

For most security problems faced by today's facility manager, the building openings are one of the most important aspects of security. While it is important to have a physical separation and necessary to understand the risks to the facility because of its type of construction—it is the openings, i.e. doors and windows that an intruder must pass through to access the assets. Therefore, understanding how to secure the building's openings is the next step. Openings in buildings fall into four categories. First, consider the doors. Doors are the openings for people to pass through into the facility. In addition, there are doors for automobiles, trucks and other vehicles like forklifts and pallet jacks that are transporting materials into and out of the building. Second, there are windows. In modern buildings, the windows do not open, but in some older facilities the windows do open and all window openings can be the weak point where the intruder breaks the glass and gets into the building. Third, there are openings for the heating, ventilation and air conditioning equipment. These openings are how the building is heated and cooled and an intruder may be able to penetrate the building through these types of openings. Last, there are other openings in the building that do not fall into the other three categories. These would be hatches, manholes, utility openings and other openings that are not designed for passage, but can be passed by an intruder.

This chapter is about doors. How they are made and which ones are more effective than others. This chapter also discusses the glazing and frames of windows—how they operate and which ones prevent an intruder and which ones do not. For ventilation openings, this chapter explains the advantages of locating them and how to cover them to prevent an intruder from accessing the facility through these openings. Since the ventilation openings can be compromised by the use of poisons, this chapter also discusses methods to protect the facility from those types of threats.

DOORS

There are many, many types of doors. Moreover, each type of door has its own problems. There are man doors, revolving doors, double doors, roll up doors, sliding doors, solid doors, metal doors, fire resistant doors, doors with glass in them and other types. Each has its own functions and weaknesses.

A secured door that cannot be opened for egress is a threat to the occupants. Hence, many building codes have requirements that govern the locking and release mechanisms for the doors to prevent people from becoming trapped in a burning building. Therefore, while it is important to have a robust door for a security reason, it is also important to have the required hardware to allow for an escape. More people have died because the door did not open and people were trapped than have been hurt because the door failed to protect them from an intruder.

This chapter does not talk about door locks, latches, and keys to control who goes through the door; that information is presented in the next chapter. What sort of opening do you have in your facility? What are the threats and how can you protect yourself from them? This information begins with the next paragraph: door types.

Personal Man Doors

A door for one person to pass through at a time is called a personal man door. It is the most common of doors and they are used for exterior as well as interior doors in commercial, institutional, educational and residential buildings. The personal door has hinges on one side and a latch on the other. There are many types of personal doors and making the door secure is an important element of security. The heavier the door, the more secure it will be. The door is an assembly that sits in a frame. The frame is then mounted into the structure. The sides of the frame are the jamb, the top of the frame is the header and at the bottom of the door, if it is an exterior door, has a threshold. See Figure 5-1 (Door Assembly).

A door that opens into the space to be protected is more secure than one that opens out because the door has what are called "stops" that hold the door. Pulling a door open does not allow for the advantage of the "stops." Another advantage of having to door open into the space to be protected is that the hinges are located on the inside and are not accessible to intruders. The door hinges are two plates with a pin through them. See Figure 5-2 Door Hinge. For extra security, the hinge pins can be welded into place.

Understanding the terms for how the door swings is helpful in understanding how the door operates. The hardware for a door varies with which way the door swings. Later in this book, sophisticated electronic locks and position sensing will be addressed. For vendors working on existing doors and

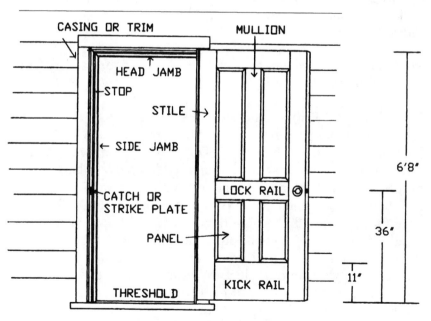

Figure 5-1. Door assembly diagram.

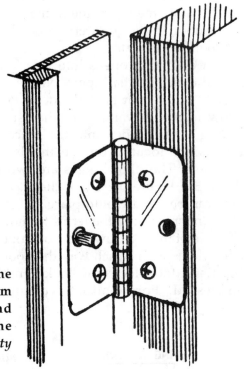

Figure 5-2. Door hinges. The pin prevents an intruder from pulling the hinges and opening the door from the back side. (*Source: Community Police Dept.*)

instrumentation systems, understanding the door nomenclature can be important.

Left Hand and Right Hand Door Explained

A swinging personal door is either a "right hand" door or a "left hand" door. See Figure 5-3 Door swing notation. The door swing can best be described this way. With your back to the hinges, a door that opens to the left with the left hand is called a "left hand" door and with your back, or butt, to the hinges a door that opens with the right hand is called a "right hand" door. In addition, there is nomenclature called a "right hand, reverse" door. This is the same as a left hand door except for the location where the keys fit into the lock. With the back to the hinges and the door opening on the right hand the door is sometimes called a "left hand, reverse" door. The "reverse" term has to do with the position of the locking/keying, which is explained in Chapter 6.

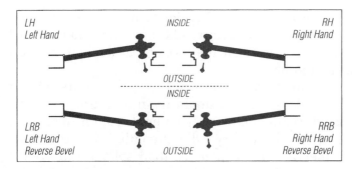

Figure 5-3. Door swing notation. (*Dictionary of Architectural Handbook*)

Doors come in varying widths, heights and thicknesses. Door width varies from two feet wide up to 3 feet 6 inches. Double doors can be wider. Door height is commonly 6 feet 8 inches with some doors available in 7 foot height. Special doors of higher height may be purchased. Finally, door thicknesses are usually 1 and 3/8 inches up to 1 and 7/8 inches. Figure 5-4 is an example of a double door.

Figure 5-4. Double door. *Photo by Robert Reid, 2004.*

If a door is heavier and wider then the door will need more hinges to hold it. For security, the minimum number of hinges should be three, one near the top, one at the bottom and one in the in the middle.

The weakest doors are hollow core doors and are not recommended as exterior doors or in a security setting. These types of doors are hollow or have light foam insulation in them and hence they have very little strength. These doors are used because they are less expensive than solid wood doors. Some doors that qualify as solid wood doors are made from a combination of wood and plastic material fabricated to look like wood or painted wood.

Steel Doors

A door that is stronger than a wood door is the steel door. Some solid wood doors are clad with steel to make them stronger. The first is a door made from steel and welded onto a frame that becomes the leaf and the second is a solid wood door with a steel cover. While steel doors can be installed with wood frames, doors with metal frames more secure than doors with wood frames.

Security forces trying to force a steel high security door with three hinges in a metal frame were not able to break the lock and finally had to break down the door from the hinge side. The screws holding the door to the hinges was what ultimately failed. It took the security team 20 minutes to break down the door using sledgehammers.

Fire Resistance

Doors are carefully measured for their fire resistance characteristics. To do this the door manufacturers furnish a complete door assembly including the frame, test it with fire on one side, and measure the heat on the other. Doors can be rated for times from 20 minutes up to three hours. In general, a higher rated fire door will be more secure than one that is not fire rated. Doors with a fire rating usually have steel frames. However, not all doors are fire rated. So it is important for the facility to know whether their doors are fire rated or not. The building codes sometimes determine where the fire rated doors have to be installed.

Doors with Visibility

Doors can be built with windows in them, although the size of the window is limited in doors that have a fire rating. Also, in fire doors the glazing (glass) in the window has certain criteria. For security, doors with windows are not recommended because the glass can be broken allowing an intruder to reach inside and unlock the door. Doors with windows that are high up on the door, and make it impossible to reach through to the latch, are better from a security standpoint. However, fire doors have size and shape limitations on the window that overrule the security requirements. Windows in doors are called "lights" because they let in light. In addition, some doors come with what are called sidelights. Sidelights are windows located next to the door. Again, security does not recommend lights because the glass can be broken and the latch mechanism reached from the outside.

Some doors that have windows are protected by bars. Other doors with windows are protected through the use of opaque glass. The opaque glass allows light in but not vision. The glazing or clear material in the window can be stronger than simple annealed glass. More information on glass for doors is explained later in this chapter with window glazing.

Security doors can have a viewer installed in them to allow the person to look through the door and see who is on the other side. Viewers are easy to install, even in a metal clad door and are a good security measure if there is no other way to determine if someone is on the other side of the door. However, for many doors it is not necessary to determine who is on the other side of the door. Doors in warehouses or doors that are protected by perimeter security would not need a light or peephole since the person knocking would already have been screened.

OTHER TYPES OF PERSONAL DOORS

While the personal man door is the most common door and the easiest to upgrade for higher security by improving the locks and hardware, other types of doors are encountered by the security professional.

Dutch Door

The Dutch door is really two doors in one opening, an upper door and a lower door. This type of door is usually used where parts or elements are often passed through the upper door when the facility is open for business. Often the lower door has a small shelf on it for laying small items like tools or gloves. The shelf also serves as a flat place for signing documents. This type of door is often seen in tool cribs and parts warehouses. The Dutch door is more vulnerable from a security standpoint than the door of a single piece, but its advantage is that it does not require opening and closing as often as a full door would in the same setting. The Dutch door can be secured with locksets as explained in the next chapter. Each piece of the door, the upper and lower can have its own lockset, or latches can latch the upper door to the lower and the lockset be placed in the upper or lower portion. A double lockset in a Dutch door is more secure than the single lockset with latches.

Double Doors

Facilities that have many people in them on occasion, like gymnasiums, auditoriums, and movie theaters are required by the fire code to have larger openings in order to allow people to evacuate quickly in the event of a fire or panic. These areas are called "places of assembly." The double door can have a post between each leaf of the swinging doors. If there is a post it is called a "mullion." The mullion improves the security of the door, giving each leaf something solid to attach to, but the mullion reduces the size of the opening and reduces the credit that can be taken for the door for meeting the fire code requirement. Double doors for "places of assembly" also have what is called panic hardware attached. Panic hardware is a bar across the door that releases the latch. The panic hardware is for a large press of people against the door in the event of a panic. Even if the person could not operate the latch, the press of the person against the door forces the latch open.

Figure 5-5. Sliding glass door. (*Photo by Robert Reid, 2004*)

Sliding Doors

In addition to swinging doors, another type of door is the sliding door. The advantage of the sliding door is that it does not require room for the door swing like the swinging door does. Photograph Figure 5-5 shows a commercial sliding door assembly. Sliding doors are easier to automate than swinging doors. The automated swinging door must have the operator set further back to allow room for the door's swing. The sliding door is most often encountered in commercial facilities like grocery stores, hotels and similar institutions. The other locations where sliding doors are commonly encountered are in a residence. Sliding doors usually have large windows or are entirely glass to allow visibility prior to opening and closing. Instead of hinges, sliding doors open and close on rollers either in the top, where the door hangs from the rollers, or at the bottom, where the door rides on the rollers. The advantage of the sliding door having rollers in the top is that there is no need for a threshold at the bottom and this type of door would allow rolling carts through the entry.

The important element of security for sliding doors is to make sure the intruder cannot lift the door off the rollers.

Revolving Doors

Another type of door encountered by security personnel is the revolving door. The advantage of the revolving door is that is works effectively to control energy losses, which are a problem with doors that open and close often. The revolving door is more energy efficient than other types of doors. The revolving door requires tempered glass so that a person can see others in the revolving portion and two persons will not try to push the door so that it operates in opposite directions. Revolving doors are expensive and require heavy traffic to be more effective than sliding doors. The other problem with revolving doors is that they won't accommodate any rolling equipment like suitcases or carts. Also, because of the difficulty for a person in a wheel chair, the revolving door will not accommodate persons with disabilities. As a result, many revolving doors are accompanied by personal doors on one side or the other.

Air Lock Doors

For energy control, an entryway is often installed with a pair of doors, which are located far enough apart that one is closed when the other one is open. The purpose of having double doors is to prevent the building's heated or cooled air from escaping and driving up the energy bills. In some cases the double openings provide for noise control, for example in a court building, an airlock separates the court from the corridor preventing conversations in the corridor from disturbing the court proceedings on the inside. Since these types doors are used primarily for energy or noise control, from a security standpoint, only one of them needs to be the door capable of being secured.

Some facilities have airlock doors offset from each other. This is to prevent a blast from blowing open both sets of doors at the same time. By offsetting the inner doors from the outer doors a blast may cave in one set of doors, but not both.

DOORS OTHER THAN PERSONAL DOORS

A facility can have doors for purposes other than personal use. These include large doors that either slide or roll-up, usually for automobiles, trucks or buses. Most often facilities with these types of doors exist to perform maintenance service on these types of vehicles, but there are also facilities where the doors open for the transfer of materials. Loading docks can be enclosed or outside with the doors opening into the facility for movement of materials. Figure 5-6 is an example of a typical loading dock.

Roll-up Doors

The roll-up type door is made from slats that are each three to four inches wide and are joined to the next slat by a groove. This allows these types of doors to roll up on a spindle above the header. Each end of the door rides in a track. The doors can be controlled with a motor that opens and closes them or they can have a chain and gear drive that is manually operated. When these doors are down, they are secured by locking the ends of the door to the track or a plate on the bottom can be locked to a notch cut into the floor. These doors are relatively heavy and, if they

Figure 5-6. Loading dock with roll-up doors. (*Photo by Robert Reid, 2004*)

have a gear drive, they cannot be opened from the outside by hand. Another security problem with this type of door is that if it is too wide, an intruder (or the wind) can bow the slats enough in the middle that one end of the door jumps out of the track. Then the intruder can simply push his way past the free end of the hanging door. Once inside, if the door is not locked, the intruder can either operate the drive or manually operate the chain hoist and open the door from the inside. These doors can be from 8 feet to 30 feet wide and up to 20 feet high.

Large Sliding Doors
The same type of sliding doors for personnel can also be used for large doors for trucks and even airplanes in the extreme application. These doors either hang or are on rollers. Depending upon their size, they can be manually or mechanically operated. For the bigger doors, a power assist is helpful in opening and closing the doors. Big sliding doors overlap. For sliding doors, it is possible to push the doors apart at the bottom and an intruder can slip in between the leaves. Once inside the intruder can cut

the locks and open the door. Most of these large sliding doors are hanging doors because it leaves the base free of objects that would obstruct wheeled traffic. Some large sliding doors have rollers on the bottom but if these rollers require a track, then a gap is created that can make it difficult for tracked vehicles like fork lifts, pallet jacks or other machines on rollers to cross the threshold.

Locking doors is explained in the next chapter on locks, locking mechanisms and keys.

WINDOWS

Another type of openings in buildings would be the windows. Since the preferred method for an intruder is the door, more effort toward security is dedicated to protection of the door than the window. However, since an intruder may be able to access a window unobserved when the door is observed, the intruder may choose the window as an access point instead of the door.

For most commercial and institutional buildings, the widows allow light into the building. The heating, ventilating and air conditioning systems are designed in such a way that the windows are kept closed. In some facilities, the windows may be capable of being opened but in these instances, the hardware for windows only works from the inside, thus, making it difficult for the intruder to open a window without breaking it.

From a security standpoint, a facility with small windows has an advantage over a facility with large windows. Another factor is the accessibility of the windows. Windows that are high up on the structure and difficult to reach from the outside are more secure than windows on the ground.

Hotel security recommends a traveler select rooms on the second and third floors. These rooms are high enough off the ground to discourage intruders while still low enough to jump out of the window or quickly evacuate down the stairs if the building needs to be quickly evacuated.

Military manuals recommend a facility have windows with a tapered sill on the outside. This would prevent and object thrown from the ground (like a Molotov cocktail or a grenade) from resting on the sill, if thrown. In a similar manner, the military favors narrow windows over wide ones. See Figure 5-7 and Figure 5-8 typical favorable window types.

In the previous chapter, a general discussion of glass and glazing was presented as a part of the discussion on building materials. The essential portion of that explanation is repeated here.

Glazing and Glass Types

The material in the window that lets in light is called glazing. Most window glazing is annealed glass or float glass. This material is brittle and easy to break and when shattered splinters into razor sharp shards. In a bomb blast, flying glass accounts for about 85% of the injuries. The thicker the glass the harder it is to break. Some windows are made from two layers of

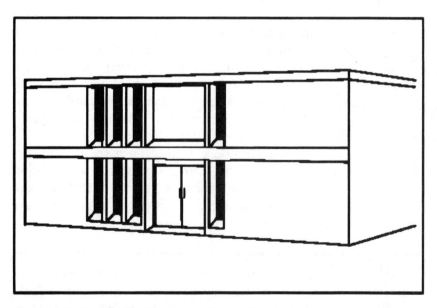

Figure 5-7. Favorable window types. Windows that are narrow are favored over wide windows. (*Source: Department of the Army Technical Manual FM 3-19.30*)

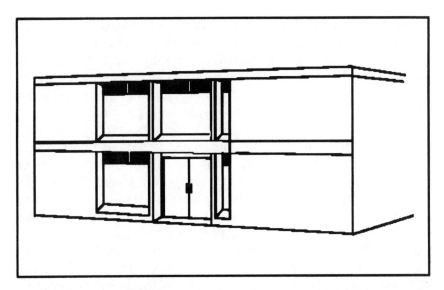

Figure 5-8. Favorable window types. Narrow windows high up on the structure are preferred to wide windows easily accessible from the ground. *Source: Department of the Army Technical Manual FM3-19.30)*

annealed glass separated by an air space. This design is primarily for heat control. It is not more effective from a security standpoint.

Security upgrades for windows with annealed glass can include application of a plastic layer on either the inner or outer surface that holds the glass if hit from a blast or with a weapon. These plastics must extend into the frame since merely coating the exposed glass would leave a weak point at the glass to frame interface. Some facilities have been installing a bar across the glass at mid-point to catch the glass after treating it with plastic to prevent its shattering. The bar is designed to catch the glass when it breaks out of the frame. Other facilities have been treating this problem with heavy drapes or curtains, but of course, this requires the drapes remain closed. Window blinds, anchored at the top and bottom are also a temporary fix.

A second type of glazing material is tempered glass. Tempered glass is required on doors and windows that can be physically touched by the public in commercial and institutional buildings. Tempered glass has been treated in a process that prevents the creation of razor like shards when breaking. Instead, the glass shatters into smaller cube like pieces. This glass can still behave like a projectile if shattered in a bomb blast, but does not splinter into the shards. Therefore, tempered glass for windows would be better from a security standpoint than annealed glass. However, this type of glass is substantially more expensive.

A third type of glazing for windows is polycarbonate materials or clear plastics. Polycarbonate material is tougher than annealed glass and depending upon the chemical makeup can be either stronger or weaker than tempered glass. One of the problems with polycarbonate materials is its weathering capability. These materials scratch easily and sometimes discolor into yellow after long hours of exposure to ultraviolet light.

Finally, window glazing can be a combination of two or more of the types. Safety glass is annealed glass that has been adhered to a plastic material that prevents the glass from coming loose. A good example in everyday practice of these types of glass is in a automobile windshield. The windshield is laminate glass and is designed to prevent an object from penetrating its layers. The side and rear windows are tempered glass and when shattered the glass falls out freely.

Bulletproof Glass

For tougher glazing, thicker materials can be used. Both the Underwriters Laboratories and the National Institute of Justice have rated standards for bulletproof glazing. These standards are rated by level. For bulletproof glass or glazing, the threat must assess the weapon for which the bulletproof glass is designed. Also with all bulletproof windows, the frame holding the glass must also be capable of resisting the impact. It is not much good if the glass pops out of the window with the first shot.

Tables 5-1 and 5-2 show the resistance level ratings from the two standards.

All three types of glazing can be colored or treated to allow light but to obscure visibility. This is sometimes seen as smoked glass, (gray colored) or frosted. This treated window glazing can improve security if the objective is to obstruct the view of intruders but in other applications, having a clear line of sight may be more important. As was discussed in Chapter 3 on

Table 5-1. UL 752 Ratings of Bullet Resistant Materials

Ratings	Ammunition	Grain's	Weight	Velocity		Number of Shots
				Min/Max (FPS)	MPS	
Level 1	9mm Full Metal Copper Jacket with Lead Core	124	8.0	1175/1293	358	3
Level 2	357 Magnum Jacketed Lead Soft Point	158	10.2	1250/1375	385	3
Level 3	.44 Magnum Lead Semi-Wadcutter Gas Checked	240	15.6	1350/1485	411	3
Level 4	7.62mm rifle Lead Core Full Metal Copper Jacket, Military Ball (.308 caliber)	180	11.7	2540/2794	774	1
Level 5	.30 Caliber Rifle Lead Core Soft Point (.30-06 caliber)	150	9.7	2750/3025	838	1
Level 6	9mm Full Metal Copper Jacket with Lead Core	124	8.0	1400/1540	427	5
Level 7	5.56mm Rifle Full Metal Copper Jacket with Lead Core (.223 caliber)	55	3.56	3080/3388	939	5
Level 8	.62mm Rifle Lead Core Full Metal Copper Jacket, Military Ball (.308 caliber)	150	9.7	2750/3025	838	5

Table 5-2 NU Standard 0801.01 Ballistic Resistant Protective Materials

Armor Type	Test Ammunition	Nominal Bullet Mass	Suggested Barrel Length	Required Bullet Velocity	Required Hits Per Armor Specimen	Permitted Penetrations
I	22 LRHV LEAD	2.6g 40gr	15 to 16.5 cm 6 to 6.5 in	320 ± 12 m/s 1050 40 ft/s	5	0
	38 Special RN Lead	10.2g 158gr	15 to 16.5 cm 6 to 6.5 in	259 15 m/s 850 50 ft/s	5	0
II-A	357 Magnum JSP	10.2g 158gr	10 to 12 cm 4 to 4.75 in	381± 15 m/s 1250 ± 50 ft/s	5	0
	9 mm FMJ	8.0g 124gr	10 to 12 cm 4 to 4.75 in	332 ± 12 m/s 1090 40 ft/s	5	0
II	357 Magnum ISP	10.2g 158gr	15 to 16.5 cm 6 to 6.5 in	425 15 m/s 1395 50 ft/s	5	0
	9 mm FMJ	8.0g 124gr	10 to 12 cm 4 to 4.75 in	358 ± 12 m/s 1175 40 ft/s	5	0

Table 5-2 NU Standard 0801.01 Ballistic Resistant Protective Materials (*Cont'd*)

III-A	44 Magnum Lead SWC Gas Checked	15.55g 240gr	14 to 16 cm 5.5 to 6.25 in	426 15 m/s 1400 50 ft/s	5	0
	9 mm FMJ	8.0g 124gr	24 to 26 cm 9.5 to 10.25 in	426 ± 15 m/s 1400±50 ft/s	5	0
III	7.62 mm 308 Winchester FMJ	9.7g 150gr	56 cm 22in	838±15 m/s 2750±50 ft/s	5	0
IV	30-06 AP	10.89 166gr	56 cm 22in	868±15 m/s 2850±50 ft/s	1	0

AP - Armor Piercing
FMJ - Full Metal Jacket
JSP - Jacketed Soft Point
g- grains gr- grams
m/s meters per second

LRHV - Long Rifle High Velocity
RN - Round Nose
SWC - Semi-Wadcutter
Cm - centimeters
In - inches

physical separation, lighting can play an important factor with glazing. Light inside on a dark night means it is easy for the intruders to see in, while a dark room with good outside lighting means it is difficult for the intruders to see in while it is easier for the occupants to see out.

For all windows, the frame that holds the glass can be metal, wood, or plastic. In some cases, the glass can be tougher than the frame. The frame is attached firmly to the structure to prevent entry by removing the window with the glass intact.

For windows that do open, hardware for the opening and closing mechanism varies greatly. Louvered windows open with a crank and sliding windows slide either vertically or horizontally, usually without rollers. Hardware to lock and unlock the window usually holds a part of the window frame with either a stop or a catch. Security officers recommend that sliding windows be protected with screens and the use of dowels or a block of wood to prevent the pane from sliding if the catch is defeated.

VENTILATION AND AIR CONDITIONING OPENINGS

A third type of opening is the opening for the building heating and cooling system. Today, buildings are so well insulated that it becomes necessary to provide openings for the air conditioning equipment to allow air in and out. In order to understand how to secure the air conditioning openings, it is necessary to understand a little bit about the air conditioning system itself.

Most air conditioning, heating and cooling systems are a loop design with a small portion of raw outside air coming into the loop and most of the air being recycled. Usually a fan moves the air through ducts installed above the false ceiling and in the walls. In some cases ducts are run in the floors, but, most of the time, the duct runs above the ceiling.

Depending upon how the facility is designed, there are large fans to move the conditioned air through the ducts to the rooms. The fans are usually combined with filters to remove dust from the air and heating and cooling coils that heat and cool the air.

The coils look like large car radiators. For many buildings the fans, filters and coils are combined into one unit called an air handling unit. A building may have several air-handling units and depending upon the building design the units can be placed separately or close together. In addition, it is necessary to exhaust some air as well as provide for fresh air intakes. Exhausts include air from bathrooms, cooking facilities, medical areas, and other areas where this air cannot be recycled.

For security protection, the building openings are usually covered with grills or louvers. In most cases, the devices are adequate to prevent entry from all but the most ardent intruder. The work necessary to penetrate the grills or louvers is much more difficult than a doorway or window. As such, unless the door and window openings have been significantly improved, the HVC openings are low on the intruder's access priority list.

However, some designs have the air conditioning louvers located in such a way that they are not visible to an outside observer. When this is the case, intruders may try to breach the building through these openings by cutting the mesh or louvers because they can perform this activity unobserved. Figure 5-9 shows the louvers on the side of an office building.

Usually, the building intake ducts are located on an outside wall near the air-handling units. For some buildings the intakes are located on the roof and for some facilities they are underground in the basement. Sometimes the grill work is horizontal at ground level covered with a grill. For safety, the grills are located away from parking areas to prevent automobile or truck exhaust from being drawn into the building.

Finally, some designs have the amount of fresh air regulated with louvers that open and close depending upon the system demand or time of day. These controlling louvers are usually at or near the air handling unit and are called variable blade dampers. Sometimes these dampers are located at the intake. The control of the dampers is as complex as the air conditioning system designer desires. Sometimes the dampers open based upon pressure, or time of day, or temperature settings or other factors. The dampers can be controlled automatically or manually. Some security designs allow the security forces to close the dampers manually but this is only a short term measure and

Figure 5-9. Ventilation opening for an office building. While many building ventilation openings are for the roof, or below ground with a grille, some facilities continue to have ventilation openings on the side of the building. (*Photo by Robert Reid, 204*)

rapidly would lead to a shut down of the air handlers. For modern buildings, with lights and computers as heat sources, a building can quickly overheat if the intake or exhaust louvers are closed.

Poisons or other Contaminants

A more likely threat to a facility through the air conditioning system is the possibility of an intruder injecting a chemical or biological poison into the building system through the fresh air intakes. This type of attack is more likely if the louvers are accessible than if they are difficult to access. For this reason, many air handler intakes are located on the roofs or high up on the side of the building. In addition, the design of air conditioning systems is such that if smoke or other particulate matter enters the system the fire protection system will close the dampers and shut down the air handling fans, which would help to mitigate the threat.

For those facilities where the air intakes are accessible, some facility managers have modified the structure to elevate the air intake above ground level by extending the duct or a structure to hold the duct, up to the roof level.

MISCELLANEOUS OPENINGS

The final class of building openings is one that is not covered by the other three previously mentioned classes. These include hatches, skylights, holes and other miscellaneous openings that are not common points of entry and are certainly not designed for human entry.

Utility Hatches

Many buildings have utilities that are provided through ducts or pipes and in some of them a hatch is provided to allow maintenance personnel to have access to the utilities when it is necessary to work on them. These openings include electric power manholes, telecommunications wiring conduits or chases, plumbing, piping and drainage. These openings are designed for the utility and not for personal access. Usually they are secured with locks and latches similar to a door. Unless the openings between the area accessed by the hatch and the facility are large enough to allow a person to fit through them, they can be eliminated as a point of access from the risk assessment. A sewer manhole or electrical duct bank manhole is usually accessed by a hatch, but if the pipes or electrical ducts are too small for a person to crawl through, these hatches don't allow access to the facility.

There is one element that should not be overlooked, however, and that would be a utility tunnel that has an inspection access along with it. This could be the case in a facility where heat, cooling water, or power is provided by a utility tunnel that runs underground from a main distribution to the facility. In this case, unless access to the utility tunnels is also controlled, it may be desirable to install a door, grille, or other barrier to prevent intruder access through this route.

Roof Hatches and Skylights

The final category of building openings includes hatches for roof access and skylights. The former is provided in facilities to allow maintenance personnel to get onto the roof, usually by a ladder or sometimes a narrow stair. These should be secured from the inside by a padlock or other mechanism. Some roofing hatches come with a predesigned locking mechanism.

Skylights are openings in the roof that are similar to windows. They are not designed to be opened, only to allow light. Skylights can be a weak point depending upon the material making up the sky light, which is usually either plastic or glass. These skylights can improve security by installing grilles or bars or improving the skylight material by using tempered glass, wire glass or polycarbonate.

CONCLUSION

Once the building's physical openings are understood and the threats to the openings are known, the next step to improving security lies in understanding the locking mechanisms that are used to secure the openings. The next chapter covers the non-electronic locking mechanisms. These are locks, hardware and keys.

RESOURCES

The Door Hardware Institute; 14150 Newbrook Drive; Suite 200; Chantilly, VA 20151; www.dhi.org

The American Association of Automatic Door Manufacturers (AAADM); 1300 Summer Avenue; Cleveland, OH 44115; 215-441-7233; www.aaadm.com

The National Fire Protection Association, P.O. Box 9101; 1 Batterymarch Park, Quincy, MA 02269-9101; 617-770-3000; www.nfpa.org

Chapter 6

Access Hardware: Mechanical Locks, Latches, Keys

Now that the physical part of the opening is understood, the next step is understanding what types of locks and latches can be used to secure the opening. This chapter discusses those elements that are mechanical in nature. No electricity or electronic power is required. Doors and other openings that are protected with electric or electronic systems are discussed in Chapter 7. But a clear understanding of the mechanical locks and latches is beneficial before trying to understand the more complex electrical systems.

Since the focus of personnel access is through doors, this chapter will be primarily devoted the different types of door locks and latches and how to secure them. This chapter also explains the use of keys, how to control keys in a large facility, and explains how a good key control policy is prepared.

LOCKS AND LATCHES

There are many types of door locks and latches and they all perform the same essential function, to secure the door in a closed position. Since the technology of key and key making has advanced, the sophistication of locking doors has increased at the same pace. Mechanical locks and latches do not require electrical power and can therefore be installed in any installation. The doors have a strike and a latch. A knob or lever is used to retract the latch from the strike pocket. In addition, a lock is used to prevent the knob from drawing the latch out of the strike pocket. Not all personnel doors have a latch and strike but these are the most common. Other types like the crossbar may be more secure but they are not as reliable.

The Strike and Strike Pocket

Modern hinged doors have a "latch" and a "strike." The latch and strike work together to hold the door closed. The jamb on the side opposite the hinges is called the strike. If the door has a latch, the part that it latches to on the jamb is called the strike pocket. The strength and security of the door is a function of how tough the latch and strike pocket are and how well they are fitted to the frame. The size and shape of the strike pocket and its strength are a part of the overall strength of a door. Figure 6-1 shows a strike and strike pocket. The gap between the door and the jamb on the side with the doorknob when the door is closed is called the reveal. The smaller the reveal the more secure the door.

For doors, the strike pocket can simply be a hole cut into the frame, but this wears out quickly on a door with a wooden jamb, therefore a metal plate is screwed into the jamb to protect the strike pocket. This metal plate is usually called the strike plate. It is rounded on the outside edge which allows the latch to ride up, over the plate and then spring into the hole. The strike plate adds strength to the latching mechanism by spreading the area of force if an intruder tries to force the door open. Some strike plates and strike pockets are larger than others and have more or longer screws. Simply removing the existing strike plate screws and inserting longer screws can improve the strength of a door.

The Spring Latch

The next element of the door that works with the strike pocket is the latch. A latch is a rounded or solid bar on a spring embedded in the door. As the door closes the latch rides up on the strike plate until it reaches the strike pocket then a spring on the latch pushes the bar into the strike pocket. With the bar extended into the strike pocket, the latch holds the door closed. Until the latch is retracted the door will remain closed. There are several shapes of latches but the most common is a round tube that fits into a round hole drilled into the edge of the door. This round tube is called a cylinder latch.

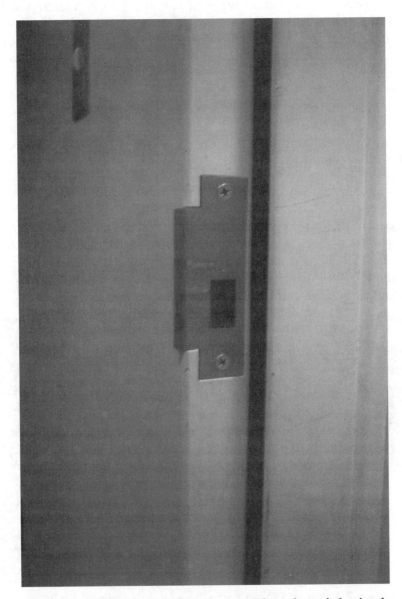

Figure 6-1. Strike and Strike Pocket. The edge of the jamb where the door closes is called the "strike." The metal hole in the door that catches the latch is called the "strike pocket." (*Photo by Robert Reid, 2004***)**

There are two main types of latches, the plain latch and the dead latch. The latch bolt that includes a dead latch is more secure than the plain latch because it is more difficult to force. The dead latch does not go into the strike pocket and prevents the latch from being extracted by locking the latch in the extended position when the door is latched. The dead latch reduces the chance of entry from pushing in a thin card or flexible piece of metal and pushing the latch back (a process called loiding). Figure 6-2 shows a latch on a cylinder lock with the dead latch pushed back. See sidebar on forcing a door for explanation of loiding. The plain latch is normally not used in security applications as it is too easy to force.

The Doorknob, Lever or Bar

Doors that have a latch need a device to retract the latch. This is usually a doorknob or in some cases it is a lever or bar. See Figure 6-3. By turning the knob or pushing down on the lever or bar, the latch is retracted and the door is free to open.

Figure 6-2. Latch and Dead Latch. The latch is the portion of the door hardware that catches in the strike pocket and holds the door closed. A "dead" latch is that part of the latch that locks the latch in place. The dead latch is forced back by the strike pocket. (*Photo by Robert Reid, 2004*)

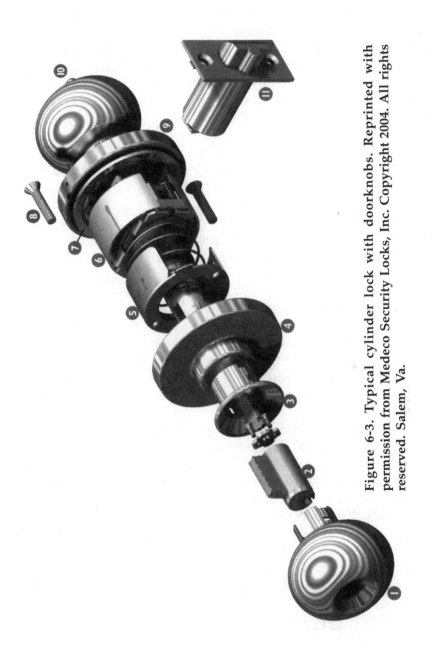

Figure 6-3. Typical cylinder lock with doorknobs. Reprinted with permission from Medeco Security Locks, Inc. Copyright 2004. All rights reserved. Salem, Va.

There are many types of knobs and latches, ones that work on both sides of the door and ones that work on only one side. Also, there are many different manufacturers of knobs, latches and strikes. The strength of the assembly is a function of the size of the latch and how strong the latch is within the door and how strong the strike is within the jamb. Doors that are kicked in usually fail because the strike pocket tears out of the frame. For higher security, a bigger strike with more screws and longer screws that extend past the jamb and into the structure increase the strength of the door lock.

On the latch side, the stronger and thicker the door, and the size and method of inserting the latch into the door make the door stronger.

These two assemblies, the latch and strike, must work together to obtain a strong door and usually the door hardware manufacturer specifies both the latch and the strike as an assembly.

Occasionally the door latch and strike is so strong that defeating the door is attempted by breaking out the hinges on the opposite side. Again, for strong doors the heavier the hinges and more and longer the screws make the door more secure.

There are four types of assemblies that retract the latch. The first and most common is the knob. Second is the thumb latch. The thumb latch is common on some residential doors, See Figure 6-4 for an example.

The third type is called the lever (Figure 6-5). The lever is sometimes required for disabled persons who would have difficulty gripping a knob or thumb latch. The final type is the bar (Figure 6-6). The bar is a hardware assembly that extends all the way across the door on the inside of doors that open out. Sometimes this bar is called the panic bar or panic hardware. It is required for rooms and buildings with a large population (places of assembly from Chapter 5). The bar goes all the way across the door and is designed to open in the event a large number of people suddenly rush the door, as in a fire, and the person at the door cannot reach or turn the knob. The press of people against the door causes the latch to retract and release the door.

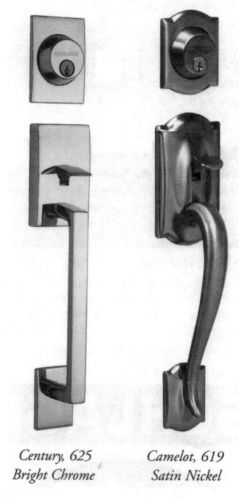

Century, 625   *Camelot, 619*
Bright Chrome   *Satin Nickel*

Figure 6-4. Typical residential thumb latch. Reprinted with permission. IR Safety and Security-Americas Copyright 2003. All Rights Reserved

Figure 6-5. Lever Latch. The lever latch works similar to a door knob but the lever has advantages because persons with limited ability to grip a doorknob or operate at thumb latch. This photo shows a door to a patient room in a hospital. (*Photo by Robert Reid, 2004***)**

Methods of Forced Entry Techniques

To force a door is different from picking the lock. Forcing a door can be accomplished by one of the following methods.

Loiding

A lock with a spring loaded latch bolt can be forced by retracting the latch bolt. This is usually accomplished by slipping a thin card, such as a credit card, or thin flexible metal bar into the opening between the door and the strike and forcing the latch to retract. Some door hardware has what is called a "dead latching" bar integral with the latch. Freeing the dead latching bar is required before the latch can be extracted. This makes forcing the latch more difficult but it can still be done. Loiding is done with thin and flexible plastic or metal cards or bands.

Jimmying

Jimmying a door is forcing it by prying the door back against the hinges or getting the door to bow or compress enough that the latch can clear the strike. Jimmying is done with pry bars. To defeat jimmying, the

Figure 6-6. Panic Hardware. This figure shows door hardware called "Panic hardware." The bar across the door will release the door. These bars are required in locations where there are large groups of people like schools or auditoriums. (*Photo by Robert Reid, 2004*)

door must fit snugly within the opening and the jambs must be rigid to prevent twisting under the force of the pry bars. A longer latch or deadbolt also reduces the possibility of jimmying the door.

Drilling

Some doors can be forced by drilling out the pin tumbler locks in the cylinder. Drilling allows the pins to retract and release the cylinder. Then a screwdriver or other flat tool inserted into the keyhole will allow the latch to be extracted by turning the cylinder. To defeat drilling, some door hardware includes what is called a slide bar that is also retracted by the key. This type of hardware is more expensive. Also, some cylinders and hardware are made from harder metals like titanium instead of the more common low strength steel or brass.

Cutting the shackle

Sometimes a padlock can be forced by cutting the shackle with bolt cutters. To defeat cutting the shackle, a larger padlock is required. Some padlocks have hardened steel shackles.

Remove or cut the hasp

Sometimes a padlock lock can be forced by cutting the hasp or removing the screws that hold the hasp in place. Then the entire assembly can be removed and the door opened. To defeat this threat, the hasp can be made of harder steels. Bolts that hold the hasp should be installed such that when the hasp is closed—it covers the bolts or screws.

LOCKS

The previous discussion has been to explain the latching mechanism of the door. It is possible to latch a door without locking it and a door can be made secure from one side by providing a latch and no method of retracting the latch from the other side. This type of lock for the door is fairly secure. It doesn't give the intruder anything to work with from the outside except the blank face of the door.

Grades of Lockset Hardware

Door hardware is manufactured in different grades. The grades define the number and types of tests the hardware must pass to meet that grade. Grades with lower numbers are higher quality than grades with higher numbers. Table 6-1 is a sample of the grade requirements for exit hardware. For security, the lower the number of the grade, the more secure it will be.

Grade 1 standards are for hardware for commercial building requirements and meet the highest security standards.

Grade 2 standards are for light commercial installations and exceed residential requirements.

Grade 3 standard door hardware meets residential requirements and generally meets standard residential security requirements.

For a facility to have the highest security, it is necessary to specify Grade 1 door hardware.

Table 6-1 Typical Test Cycles for Various Grades of Door Hardware.

Test	Grade 1	Grade 2	Grade 3
Cycles	500,000	250,000	100,000
Exit	15 lbf max. (67 N)	15 lbf max. (67N)	15 lbf max. (67N)
Outside Pull	400 lbf min. (1780 N)	400 lbf min. (1780 N)	300 lbf min. (1335 N)
Inside Pull (Grades 1 & 2)	400 lbf min. (1780 N)	400 lbf min. (1780 N)	
Push (Grades 1 & 2)	400 lbf min. (1780 N)	400 lbf min. (1780 N)	
Forces to Latch Door	4.5 lbf max. (20 N)	4.5 lbf max. (20 N)	4.5 lbf max. (20 N)

Source: Builder's Hardware Manufacturers Association. www.buildershardware.com/standards

SITUATIONS FOR LOCKING AND UNLOCKING

Locks are the devices in a door latch that prevent the latch from being extracted. Locks come in many different configurations for many different types of applications. For exterior doors, the common practice is that a key unlocks the door from the outside. While on the inside there is a lever or button which locks the door. The door can be either locked or unlocked. When the door is locked, it can be unlocked from the outside with the key. The inside doorknob will open the door without the key whether the door is locked or unlocked. For some situations the local building codes require that the door can be opened from inside without the key.

Another type of lock is one that can be locked or unlocked with the key from one side while the other side is always unlocked. This is typical of schools where a class room is locked from outside but if someone is inside, they can exit without a key.

A third type is one that is always locked and can only be opened with the key from one side. The other side is always unlocked.

For some high security applications, it may be necessary to have a door with a lock that requires a key from either side. However, this could conflict with fire code requirements to exit.

From a security standpoint it is obvious that the safety requirements for exit can conflict with the security requirements and in general the fire safety requirements will overrule the security requirements.

TYPES OF DOOR LOCKSETS

There are generally three types of locksets, the cylindrical or tubular lock, the mortise lock and the deadbolt lock.

Cylinder Lockset

The cylinder lockset is installed in a round hole cut into the door at the doorknob. See Figure 6-7 Typical Cylinder Lock assembly. The cylinder lock is easy to install and is sufficiently robust for most security applications. Because the key fits into the doorknob in these locksets, it is sometimes referred to as a "key-in-knob" lockset. Also, these locksets can come with a lever instead of a knob to make the door easier to open for someone who cannot grip the knob to turn it.

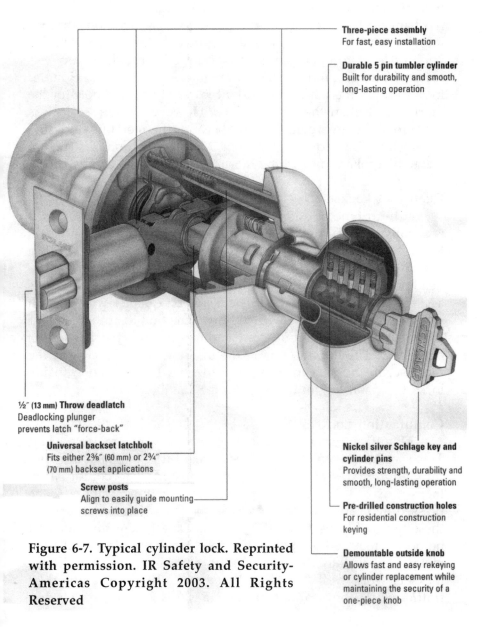

Three-piece assembly
For fast, easy installation

Durable 5 pin tumbler cylinder
Built for durability and smooth, long-lasting operation

½" (13 mm) **Throw deadlatch**
Deadlocking plunger prevents latch "force-back"

Universal backset latchbolt
Fits either 2⅜" (60 mm) or 2¾" (70 mm) backset applications

Screw posts
Align to easily guide mounting screws into place

Nickel silver Schlage key and cylinder pins
Provides strength, durability and smooth, long-lasting operation

Pre-drilled construction holes
For residential construction keying

Demountable outside knob
Allows fast and easy rekeying or cylinder replacement while maintaining the security of a one-piece knob

Figure 6-7. Typical cylinder lock. Reprinted with permission. IR Safety and Security-Americas Copyright 2003. All Rights Reserved

Mortise Lockset

The mortise lockset is a different design from the basic cylinder lock. The mortise lock is named because it fits into what is called a mortise in the door. A mortise is a rectangular shaped hole cut out from the door edge that is designed to fit the lock mechanism. The latch on these types of locks is in the box that fits into the slot cut into the door. See Figure 6-8 Mortise Lockset. The knob or lever to open the latch is attached through the door as before. In general mortise locks are more expensive than cylinder locks because there is more work necessary to cut the hole for the mortise lock than the cylinder lock. However, a mortise lock is more robust than a cylinder lock because the metal box holding the latch is embedded in the door. More of the door's structure has to be broken to force the mortise lock than the cylinder lock.

Dead Bolt Lockset

A third type of door lock is called the dead bolt lock and it is usually used in combination with one of the other two types of locks. This type of lock is called a deadbolt because the latch is a straight bolt that goes through the door and into the jamb. The dead bolt lock is usually in the door above the other lock and acts as a second latch for the door. The dead bolt can be keyed on both sides, keyed on one side with a thumbscrew on the other side, or it can have only the thumbscrew and no opening for a key on the outer side. The thumbscrew in the inside with no exterior is the most secure but it can only be utilized when someone inside locks the door. Figure 6-9 is a typical deadbolt lockset.

Combination Lockset

Manufacturers make a lockset that does not require a key. This is called the combination lockset and it is available in either the cylinder design or the mortise design. The combination lockset uses a numeric keypad requiring the numbers to be pressed in a certain order to release the latch. The advantages of these types of locksets are that are there are no keys to issue and no power required to operate the door. Some brands also allow a metal key, so that the combination can be overridden. These locksets work well for doors that have a medium amount of traffic where they cannot be tended. (A door with an electric

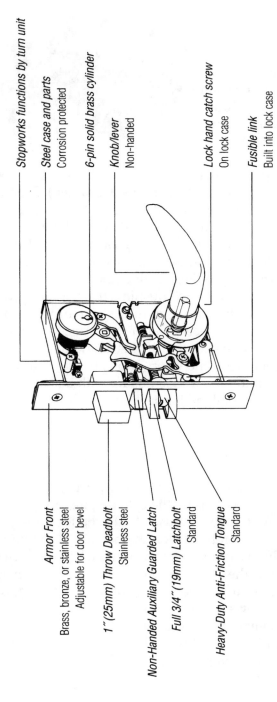

Stopworks functions by turn unit

Steel case and parts
Corrosion protected

6-pin solid brass cylinder

Knob/lever
Non-handed

Lock hand catch screw
On lock case

Fusible link
Built into lock case

Armor Front
Brass, bronze, or stainless steel
Adjustable for door bevel

1" (25mm) Throw Deadbolt
Stainless steel

Non-Handed Auxiliary Guarded Latch

Full 3/4" (19mm) Latchbolt
Standard

Heavy-Duty Anti-Friction Tongue
Standard

Figure 6-8. Mortise Lockset. Reprinted with permission. IR Safety and Security-Americas Copyright 2003. All Rights Reserved

Figure 6-9. Deadbolt lockset. Reprinted with permission from Medeco Security Locks, Inc. Copyright 2004. Salem, Va. All rights reserved.

strike would be preferred if the door was tended by a person. Electric strikes are explained in the next chapter.) People who know the combination can pass through without having to be issued a key. The combination can be changed easily. Usually these locksets fit into the same cutouts for a standard door and their latch and deadlatch mechanisms work the same as other hardware. They can be provided with a knob or a lever to open. The combination is entered by pushing the keys in sequence; sometimes the combination requires pushing two keys simultaneously. For a few seconds after the correct combination is entered the door is unlocked but if the doorknob or lever isn't turned within a moment, the door automatically relocks. These hardware sets retail for about $450 dollars with installation extra. Figure 6-10 shows a combination lockset.

KEYS AND CYLINDERS

Each manufacturer of door hardware provides keys to unlock their type of lock. Inside the cylinder are tiny pins resting against springs. As the key is pushed into the lock, the pins are raised or lowered depending upon the heights of the bumps on the keys. The cylinder is designed such that when the correct key is placed fully into the lock, the pins are at the right height to allow the cylinder to turn. The turning of the cylinder draws the latch back and allows the door to open. Figure 6-7 shows a typical key and cylinder arrangement.

These keys, pins and cylinders have become more and more complex in an effort to improve security. The more pins there are the more difficult the lock is to pick.

**Figure 6-10.
Combination lockset.
Reprinted with
permission from Kaba-
Ilco Corp. Copyright
2001 Kaba-Ilco Products
Catalog, Winston-Salem,
NC.**

LOCKPICS

A series of tools can be used to "pick" a lock without using the key. This is called lock picking. A small pointed tool is inserted into the lock and each pin is raised or lowered while turning force is applied against the lock. As each pin reaches the right height the turn can hold it against the edge of the cylinder. A good locksmith can pick a 7 pin lock in about 30 seconds. To increase security, more sophisticated cylinders with more pins can be used. One manufacturer (MEDECO) has devised a way of angling the pins making the lock more difficult to pick.

In a effort to make these locks more secure, top and bottom pins have been added as well as pins that must be aligned horizontally as well a vertically. Figure 6-11 shows a Schlage Primus High Security Key lock. Figure 6-12 shows a Medico Key and interchangeable core. There are many other types of lock manufacturers but these all work on a similar principle, a key fits into the lock, aligning the pins and allowing the lock to turn. It is a mechanical system that has been used effectively for many years.

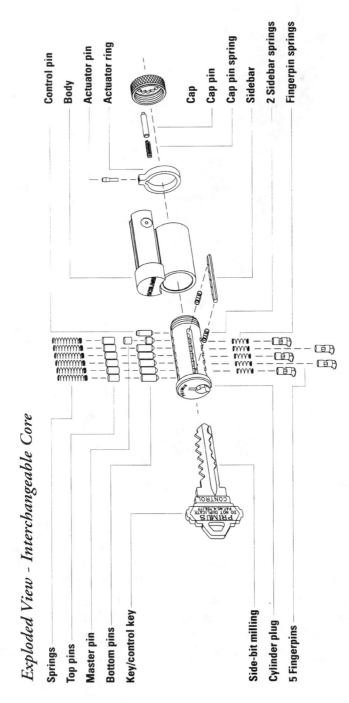

Figure 6-11. Schlage Primus high security lock. Reprinted with permission. IR Safety and Security-Americas Copyright 2003. All Rights Reserved

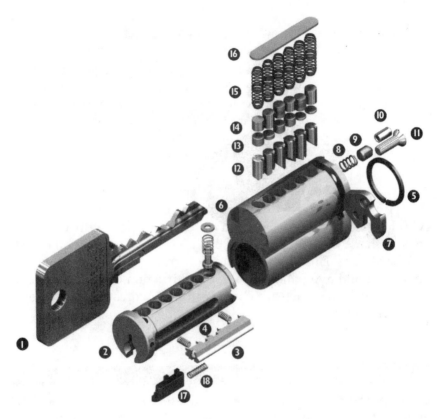

Figure 6-12. Medeco high security key and core. Reprinted with permission from Medeco Security Locks, Inc. Copyright 2004. Salem, Va. All rights reserved.

Master Key and Grand Master Key

Many of the locks have more than one pin in each vertical slot. A close examination of Figure 6-12 shows several types of pins in the cylinder. These are called top pins, master pins, bottom pins and finger pins. To provide a higher level of access and control, a lock can have, in addition to its own key, a separate key called a master key. Just as a single key will open the lock by setting pins to the correct height allowing the cylinder to turn, the master key will also raise pins to a correct height that will allow the cylinder to turn. Finally, at some facilities a third level, called a grand master key, will also open the lock.

Another way of defeating mechanical keys and cylinders.

It is possible to drill out the lock using an electric drill by drilling a hole through the cylinder at the point of the cylinder interface. Instead of using a key to raise the pins to the correct height, a drill bores through the pins and allows the cylinder to turn. It takes longer to break a lock this way and it is noisier than the lock picking method.

Higher standard door hardware comes with removable cylinders called cores. Different cores will open with different keys so that a facility can be re-keyed without changing all of the lock hardware. This is typical on a construction project where the doors will all have what are called construction cores. At the end of the job, all the cores are changed to new cores that have keys from the facility.

Types of Keys

Figure 6-13 shows a key and its typical parts. These elements are important to locksmiths and sometimes to intruders. The manufacturer of the key can be identified by the bow of the key. Knowing the bow determines which type of blank key will fit the same lock. Locksmiths use the bow to determine which type of key is needed when new keys are made. Many key manufacturers now have a policy that only certified locksmiths can re-cut keys for their locks. In this way, the number of keys and the ability to duplicate keys is somewhat restricted. Many keys have "DO NOT DUPLICATE" stamped on the bow. In general, locksmiths and key shops will not duplicate a key with this requirement. However, this would not stop an illegal intruder provided they had the ability to make or cut their own keys. Since the key manufacturer restricts the availability of some blank keys, it also restricts the ability of intruders to duplicate keys.

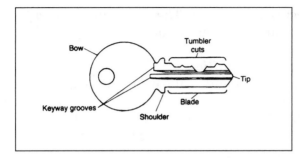

Figure 6-13. Typical key and component parts.

IMPRESSIONING A KEY

Some locksmiths and intruders can gain access by performing a technique called impressioning a key. What they do is find a blank key that is the same manufacturer as the lock they wish to defeat. Then using a powder or other material to mark the key, they put the key in the lock and attempt to turn it. Then when they extract the key, the powder is marked by the pins and they file the key at this point. By inserting the key and turning it several times and successively filing it after each time, they can create a key that will open the lock. This method takes longer than picking or drilling the lock, but with a new key that has been impressioned, access is available for whenever needed. The higher grade the lock, the more pins and the more sophisticated the lock, the more difficult it is to impression a key. Again, if the brand of the key is known, the intruder may be able to obtain a blank. Some manufacturers and locksmiths guard blank keys closely to prevent re-keying or impressioning.

Key Control

A facility should have a key control program. Key control is a program or policy designed to protect the facility's security by controlling who has access to keys and what the keys unlock. A good key control program starts with everyone who has a key understanding that their key is a part of the overall security process. Key and lock control policies are an important element of any security program. The following elements should be included in any key control policy.

Employee Responsibilities

The policy should state what the responsibilities of the employees are with regard to building locks and keys. How they can be issued keys, how they are to control their keys, what to do if a key is lost or stolen and what to do with keys at the end of their need for the key.

Many facilities mandate that if a person does not have a key to the door, they are not allowed through that door. They must obtain permission from someone higher up in the managerial chain.

The policy should indicate to the employee what action should be taken should a lock need to be changed.

The policy should indicate what the employee should do if one of the keys is lost or stolen.

Managerial Responsibilities

The policy should indicate who has the authority to issue keys. What keys are authorized? In general, whoever is responsible for the space should be responsible to determine who is issued keys. In addition the policy should consider whether keys can be loaned to others, and if so, to whom? Keys with high levels of access should be authorized by higher levels of management. In some facilities it is forbidden for someone to authorize a key for themselves.

Locksmith or Key Shop Responsibilities

The policy should indicate how keys are obtained. Who actually makes the key should be stated and what authority is required for making the keys.

Key and Lock Control Form

Many policies include a key and lock control form. This form is a device to track who has been issued keys and what keys they have been issued. It also gives an indication of who authorized the issuance of the key.

The same form can be used to request a change in locks, or cylinders, for a space where the function is changed. At the end of this chapter is a copy of a public domain key control policy.

Stolen master key costs Bend College $35,000

BEND, ORE.—When a master key to Central Oregon Community college went missing last fall, there was plenty of consternation on campus. After all, whoever had taken it had access to every classroom, every laboratory, and every office on campus.

Eventually, a temporary custodian was convicted of stealing the key, but not before locks had been changed, additional security officers had been hired and $35,000 spent on new locks, new keys and the labor to make the change.

The question that remains is whether anyone actually copied the key.

"Right on it, it says 'do not duplicate'," Gene Zinkgraf, the director of campus services, told *The Bulletin* in Bend, Oregon. But someone did try to copy the key, he said, based on the fact that a part of the key is ground down.

The former custodian has been ordered to pay the campus $34,000 in restitution.

Source—Central Oregon Community College July 2004.

Key Control Lock Box

In some facilities keys are not carried by individual employees because for each employee to carry a key for each space would make the key control very cumbersome. As a result keys are kept in a central place, called a key cabinet. The key cabinet is kept locked and when employees want to draw a key, they go to an administrative assistant familiar with the policy of who is authorized to draw keys. The person requests the key they need. The assistant issues the key, noting who requested it, what was requested, when the key was logged out and when it was returned.

This type of logging can become cumbersome and some facilities have been able to make this work easier by computerizing it and keeping all the information in a database.

Some lock vendors also provide, with their locks and keys, software programs for lock and key control.

THE DOOR CLOSER

The final hardware element for doors is a device called a "closer." See Figure 6-14. The door closer is a device that allows the door to swing open and then slowly brings the door closed. The closer is a good device for keeping doors closed and, because it slows the swing of the door at the end, it helps to preserve the life of the door hardware. Closers come for doors to swing open 180 degrees and 90 degrees and it is important to make sure that the correct closer is installed to prevent over opening the door and damaging the door, the header or the hinges.

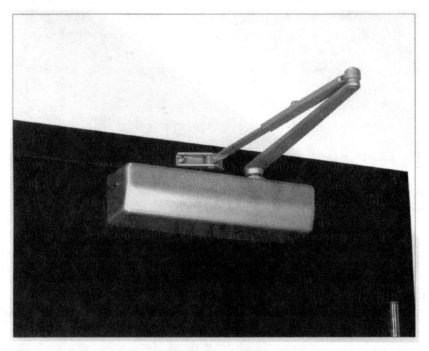

Figure 6-14. Door closer. Reprinted with permission from Corbin-Russwin Inc. Copyright 2004 All rights reserved. Monroe, NC.

CONCLUSION

Doors have mechanical hardware to close and latch the door and the hardware has locks and keys for security. Mechanical hardware is excellent and it requires no electrical power to control the doors and locks. However, new electronic hardware is being installed every day to improve security and provide better control. The next chapter provides information on electronic door keys and locks.

RESOURCES

For additional resources on this topic a number of manufacturer's offer CD ROMs of their catalogs and additional information on their web site.
Builder's Hardware
Schlage Lock Manufacturing Company www.schlagelock.com
Securitron Magnalock Corp. www.securitron.com
Medeco Lock Company www.medeco.com
Kwikset Lock Company www.kwikset.com
Kaba Ilco Lock Company www.kabailco.com
There is a very interesting article on how locks work that includes animations of how key opens a lock on the web at http://home.howstuffworks.com/lock-picking.htm

---

Many Key and Lock Control Policies can be found publicly on the Internet. The University of Nevada Reno had this lock and key control publicly posted.

UNIVERSITY OF NEVADA RENO
KEY CONTROL POLICY

PURPOSE
The purpose of this policy is to provide optimal physical security for the campus community and to protect the assets of the University of Nevada, Reno (the University)

EMPLOYEE RESPONSIBILITIES

An employee is responsible for any and all keys issued to them.

1. Employees will NOT loan or transfer their keys to any other individual.

2. Employees are not to unlock doors for others. These requests are referred to the department office or the University Police Department. Authorized card key holders are not allowed to let anyone into a building after business hours under any circumstances except for public safety personnel if required for an emergency.

3. Keys will be returned to the University Lockshop when an employee terminates or transfers within the University.

4. Doors to unoccupied rooms are to be locked when not in use.

5. Outside doors are to be locked after normal business hours.

6. Maintenance gates and chains are to be locked immediately after passing through.

7. Unauthorized people or any suspicious activities are to be reported to the University Police Department.

8. Any University keys found should be turned in to the University Police.

KEY DEFINITIONS

Building Master Keys: These keys allow access to all rooms within a given building.

Building Submaster Keys: These keys usually allow access to several areas within a building.

Card Keys: These are plastic cards coded with magnetic data which have been programmed into a central building access computer control system.

NOTE: Card Keys are becoming a part of our campus key system; therefore, the term key and card key will be interchangeable throughout this document.

Keyed Alike Keys: These keys allow access to multiple rooms within a single building.

Individual Pass/Space Keys: These keys allow access to one room or a single space.

Replacement Keys: These keys are defined as any key given to replace a lost or stolen key or card key.

Lockshop: Located in Buildings & Grounds, 1305 Evans Avenue.

Work Control Center: Central Processing Department where all requests for key services are sent for processing (ext. 8020) in Building Grounds Department, 1305 Evans Ave.

AUTHORIZATIONS

The Lockshop will have a signature card on file for each Chairperson, Dean, Assistant Vice President, or Vice President.

Building Masters and Submasters must be authorized by the Dean, Assistant Vice President, or Vice President responsible for their area ONLY.

Keyed Alike and Individual Pass/Space keys can be authorized by a Chairperson, Dean, Assistant Vice President, or Vice President.

Card Keys will only be issued with written approval by the Chairperson, Dean, Assistant Vice, President, or Vice President and the completion and acknowledgment of the mandatory training.

All key requests must be initiated on the proper key request form and submitted to the Work Control Center in advance of the date needed. Requests may take up to 10 days to process. Failure to have the proper signature authorization will result in keys not being issued and the request being returned to the requesting department.

SIGNATURE STAMP IS PROHIBITED

KEY FEES

The following key deposit policy will be used to determine costs for duplicating keys.

Administrative faculty, instructional faculty, and staff personal will NOT be required to PAY for their INITIALLY AUTHORIZED KEYS but WILL be required to PAY for additional keys replacing stolen or lost keys due to departmental negligence.

All costs associated with rekeying a department's building space will be the department's responsibility if rekeying is determined by the Assistant Vice President or Director of Facilities Services Department to be required due to departmental negligence.

Departments will be alerted and charged for the Building Master Keys after four have been requested.

Departments are required to submit an authorized key request form and pay for keys for temporary staff, students, and visiting faculty members. One key per individual and each individual shall sign for their key. Returned keys will be credited to the department.

Key fees are as follows:

Building Master Key	$250.00
Submaster Key	$ 50.00
Individual Pass/Space Key	$ 10.00
Keyed Alike	$ 10.00
Card Key	$ 10.00

Broken or worn out keys will be replaced at NO CHARGE, but must be brought to the Lockshop in person; or send an authorized Department representative with a signed and dated key request from requesting the key to be replaced. The new replacement key will then be stamped exactly like the broken or worn out key. Call the Work Control Center in advance to schedule an appointment with the Lockshop for this service. Return broken and worn out keys to the Work Control Center for disposal and refund purposes.

KEY TRANSFERS

For security and personal safety reasons, the transfer of keys from Department personnel, faculty members, students, and University staff is PROHIBTED.

KEY CONTROL AND INVENTORY

Each Department will be responsible for keys issued within the Department. It is recommended that each Department maintain their own internal, written inventory of keys and issue only to personnel that need to have access to your facility. An inventory of all Master/ Submaster issued on campus is maintained in the Work Control Center. If a copy of your department's key inventory is needed, contact the Work Control Center, in writing, for this information.

Requests for key cards shall come from the Departments to the Work Control Center. All card keys will be issued through the Departments from which they have been requested after they have been programmed by the Lockshop.

When a Department wishes to issue card keys to students, the Department must verify that the student is actually enrolled at the University.

RECOMMENDED AUTHORIZED KEY HOLDERS

Dean/Dean's Secretary (Building Master), Department Chair/ Department Secretary (Submaster), Full Time Faculty and Staff (Pass/ Space Key, Classroom Key), Visiting/One-Year Appointment Faculty (Pass/Space Key, Classroom Key), Graduates/Undergraduate Students (Pass/Space Key, Classroom Key). Card Keys will be issued to those requiring access after regular business hours as approved by the proper authority.

UNAUTHORIZED DUPLICATING/REPLACING KEYS

Duplicating or replacing keys through an agency, company, or private business other than the University of Nevada, Reno, is NOT ALLOWED and is a breach of this key policy. When this violation has been discovered, the appropriate Chairperson, Den or Vice President and the University Police will be notified for appropriate action.

NOTE: The duplication or possession of any unauthorized University keys is a Misdemeanor, NRS.205.080.

TEMPORARY KEYS

Temporary keys for visiting professors, temporary employees and students may be issued. A key request form stating when the keys will be returned (1-2 per semester) and letter dated and signed by the appropriate authority must be submitted prior to the issuing of keys.

Card Keys requests for visiting professors, temporary employees and students must specify an expiration date on the key request form.

RETURNING KEYS

Prior to leaving the University, all keys MUST BE RETURNED to the Lockshop and refunds, if any, will be made. It is the responsibility of the Department to retrieve all keys from departing employees. The employee should copy all records supporting the number and type of keys returned to the Lockshop for future reference and key refund credit.

Employees failing to return their keys before leaving the University may have a financial hold placed on their final paychecks and/or final grades/transcripts. Employees transferring from one location to another within the University ARE REQUIRED to RETURN their present keys and request via the key request form, keys for their new location.

LOST OR STOLEN KEYS

All lost/stolen keys MUST BE REPORTED IMMEDIATELY to your Department, the Work Control Center, the Lockshop and University Police Department.

Use the key request form to report the lost or stolen key and a memorandum should accompany the form stating all facts about the incident. A copy of the memorandum and key request form will be sent to the University Police Department to ensure immediate concerns are addressed.

Information to be included in the memorandum when reporting the lost/stolen key is as follows:

—Name, Department, date and phone number of employee.

—Circumstances surrounding the incident and where the key(s) were lost or stolen.

—Buildings and areas which are affected by the lost or stolen key(s).

It is important to document all information regarding the incident and all details should be included. This will help in any investigation which may follow the incident.

INSTALLATION AND REPAIR OF LOCKS

All installation and repairs of door locks and mechanisms will be performed by the Facilities Services Lockshop. When remodeling or building renovation work is being performed by the University Renovation Staff, all non-university locks that are encountered will be removed at the Department expense. This will include the cost of any hardware necessary to complete the needed repairs.

CARD KEY ACCESS

The following buildings will have the perimeter security system that will require a card key to enter and exit:

1. Ansari Business Building
2. Chemistry Building
3. Church Fine Arts
4. William Raggio Building (Formerly College of Education)
5. Fleischmann Agriculture Building
6. Lombardi Recreation Building
7. Paul Laxalt Mineral Engineering
8. Paul Laxalt Mineral Research
9. Ross Hall
10. Sarah Fleischmann Building
11. School of Medicine (All buildings)

PERIMETER SECURITY-CARD KEY ACCESS POLICY

The above list of buildings has the perimeter security, card access system, which requires a card key to enter and exit after normal business hours. Exterior doors are automatically locked magnetically after a predetermined hour, which will be posted at each exterior door.

BUILDING HOURS: Each building will have its own schedule for opening and closing as determined by the appropriate Dean for each building. Generally, operating hours will be from 7:00am to 11:00pm Monday through Friday. When the building is closed the magnetic door locks will automatically energize. On Saturdays and Sundays most buildings will be scheduled closed except when special events are arranged through the University Schedule Office. In special cases some buildings will be open on Saturdays and Sundays. Heating, ventilation and air conditioning will be on a setback mode when buildings are closed except for research facilities. Facilities Services Department will adjust the temperature control when buildings are opened for special events, during their normally closed period, as requested by the Scheduling Office.

ENTRANCE/EXIT: Buildings with the perimeter security system have two primary entrances with video surveillance and card key access from the exterior of the building after regular business hours.

AUTHORIZED CARD KEY HOLDERS: It is required that each person who is authorized to be in a campus building, after business hours, be responsible to have his or her card key with them at all times. This will enable each person to exit the building for non-emergency conditions. Authorized card key holders are not allowed to let anyone into a building after business hours under any circumstances except for public safety personnel if required for an emergency.

FIRE ALARM EMERGENCIES: The perimeter security system in each building is tied to the building fire alarm system. In the event of a fire alarm or a loss of power the perimeter security system will disarm the magnetic door locks. A card key will not be required to exit the building.

NON-FIRE ALARM EMERGENCIES: In the event of a non-fire alarm emergency such as personal assault, medical emergency, chemical spill, etc. call 911 from the nearest telephone. If calling from an emergency phone located at a primary entrance identify your call to Central Dispatch as an emergency, they will transfer your call to the 911 dispatch center. Center Dispatch will send the University Police and any other required emergency services. The University Police Department will open that building to give access to other services needed.

VIDEO RECORD: A video record of those individuals entering and exiting a building will be provided at the primary entrances and will be recorded when a building is officially closed. Full time video monitoring by University personnel will not be provided.

TRAINING: Training for the use of the card key will be required. You will be required to know the following:
  Emergency Procedures
  Where the main entrances and emergency exits are located
  How to use the card key

Chapter 7

Electronic Access Hardware: Locks

Mechanical locks and latches for securing openings have worked well for many years, but many facilities with a large staff want more security with less key management. The next level in upgrading a facility's security is to utilize the electrical applications for access control and to improve the overall security picture.

Using electrical power to lock and unlock doors is more complex and more expensive than the mechanical locks discussed in the previous chapter, but when keys and re-keying are eliminated from an analysis; some of the costs can be reduced.

This explanation of electronic access control begins with the simplest form, the electronic lock, and progresses to successively more complex systems like the electric strike and electrically operated mechanical door hardware. Finally, it is important to know and understand the types of devices that control locking and unlocking electrically; which is also addressed in this chapter.

UNDERSTANDING BASIC ELECTRICAL PRINCIPLES

Fundamentally, using electricity for door control is a simple process. The most difficult part is understanding what is going on in the wires and to do this several tools are needed that are not in the common locksmith's toolbox. The most common of these tools is the electrical multi-tester. This device is used by electricians and trained locksmiths to measure the strength and direction of electrical signals. Knowing there is enough power

127

and whether it meets the requirements of the device is the first step. The next one is to bring the power to the device.

Electrically powering door hardware, or anything else covered in this book, is a matter of understanding what amount and direction of electrical flow is required. This gets a little more complex for multiple systems because the devices use power at different levels. Providing electrical power for other devices like intrusion detection and controlled circuit television systems is covered in later chapters. Since many facilities don't want or need a complete burglar alarm system and may simply want an electrically locking door that information is covered here. Intrusion Detection Systems are explained in Chapter 10: Intrusion Detection Systems.

Electricity for doors does one of three things. First magnetism is used to lock doors. These are called electromagnetic locks. Second, devices called solenoids are used to push and pull levers. A solenoid is really only a magnet with a plunger. The electricity magnetizes the plunger and it extends or retracts, pushing or pulling whatever has been attached to the plunger. The third element of electric door access control is using an electric motor to turn something. The electric motor can open automatic doors, for example, or push or pull hinged doors open or closed. The last of the electronic duties for door control is the electric reader that reads a card or palm print that gives permission to unlock or denies access by keeping the door locked. Since there are so many types of electrical devices used to unlock an electrical locking door, that topic has its own chapter following this one, Chapter 8: Electronic Lock Control.

The most simple of the electric door controls is the electromagnetic lock.

ELECTROMAGNETIC LOCKS

The electric lock doesn't modify the old mechanical hardware at all; it simply adds another layer of security by locking the door separately from the keys and mechanical systems in the previous chapter. The most common of the electric locking mechanisms is the electromagnetic door lock (Figure 7-1).

Figure 7-1. Electromagnetic Lock. This electromagnetic lock is being used in a school to hold doors open. The device can also be mounted to hold the door closed. (*Photo by Robert Reid, 2004*)

The electromagnetic lock is a flat metal plate that mounts to the door that works with a magnet that mounts to the header. There are several types of systems but the most simple is that the magnet takes electric power to hold the door closed. Magnetic locks are available in different strengths requiring from 300 pounds to 3000 pounds to break the lock. The magnetic lock system is secure provided power is applied correctly. Figure 7-2 shows an installed magnetic lock in a metal framed door.

Just the same as any other electrical device, power for the magnetic lock is provided through wires from a power supply source. The wires are usually run overhead in a false ceiling and then "dropped" to the magnetic lock at the header. Sometimes power is provided through conduit mounted on or in the ceiling and sometimes power is fed to the device through wires that run through the walls between the jamb and the structural elements. In some instances, with metal framed doors, for instance, the wires can be fed through the hollow metal of the frame. Usually the power is mounted to the frame and the metal plate in the

Figure 7-2. Installed electromagnetic lock. Reprinted with Permission from Securitron Magnalock CORP-Copyright 2004 Sparks, NV.

door does not require power. So for the magnetic lock the power does not have to be run through the door.

The power for magnetic locks depends upon the manufacturer of the device. Usually, however, the power is either 12-volt direct current or 24-volt direct current. Since most building power is alternating current a device called a transformer is needed to step down the power and convert it from AC to DC. Depending upon the number of doors that have magnetic locks this may be one transformer for the entire system or individual transformers for each lock. Power problems with magnetic locks can be the result of many factors such as low voltage, wrong current, crossed terminals and "signal noise." Signal noise can mean that the power is fluctuating.

The final element for a security manager to understand about magnetic door locks is safety. Of course the wiring must be installed according to local codes. However, while the magnetic lock keeps intruders out, unless there is a means to cancel the power, the door is locked for people inside who need to exit. As a result most magnetic locks have a power override switch to release the door from the inside, if necessary. This means an extra switch has to be installed on the wall inside the door. In another facility there may be an electrical connection between the door knob or lever on the inside and the electromagnetic lock. This way, when the knob or bar is pushed, the electromagnetic lock is released. This installation is more complex because power cables have to be run through the door to get to the knob.

Some manufacturers have devices that sense presence from a motion detector or infrared sensor from the inside and automatically unlock the door when the person tries to exit. Again these devices increase the cost of the electromagnetic door lock.

At the Benton County Justice Center in Kennewick, Washington, the court sometimes had a problem in civil cases with people "bolting" for the door when the sentence was handed down. What was happening was that when the husband or spouse was told what the alimony payment would be they would "run for it." The justice center installed magnetic locks on the doors that were controlled by the judge at the bench. If the defendant bolted, the judge could lock the doors from the bench. Because the courtroom is a "place of assembly" the lock could not stay locked. It just started a 15 second timer. But that gave the sheriff's deputies enough time to come to the courtroom from next door and detain the suspect.

This, request to exit delay, is not allowed in all building code jurisdictions so a code review needs to be carefully checked if this type of installation is planned.

The other power element the manager needs to consider for magnetic locks is what happens if the power fails. Usually, it has been determined that it is better for magnetic locks to release upon loss of power, since this would allow building occupants to exit. However, an alternative is to provide backup power from batteries or another source to hold the lock if the power fails. There is more discussion of power supplies and backup power in Chapter 13: Power Supplies. Some electromagnetic door locks are battery powered so it isn't necessary to run wiring to the door.

Finally, it is important to recognize under what conditions the door will release when a person wants to come in through the door. The terminology here is that the candidate must present a "credential" that identifies him and releases the lock. Sometimes this is managed by someone who tends the door, as in a jail or prison. The attendant looks through the window, recognizes the person, and presses a switch that unlocks the door. In another case a person may have a badge with a code written on it that they hold up to a verification device. If the badge is authorized then the lock is released. There are many ways for a candidate to present the "credential" from using a key, a badge, a thumbprint or hand or retinal scan. This topic will be covered in more detail in Chapter 8: Electronic Lock Control: Giving Users Access.

Some electromagnetic locks come with a custom designed header designed to fit into the frame. This camouflages the magnetic lock and generally makes the door opening more attractive.

A magnetic lock costs between $50 and $150 dollars, but the extra costs of providing DC power, the wiring, the release mechanism for exiting and the credential can increase the cost up to $3000 per door.

Shear Aligning Magnetic Lock

Because the magnetic lock is unattractive to some facilities there is an alternate to the basic magnetic lock. This alternative is called a "shear aligning" magnetic lock. The shear aligning magnetic lock fits inside the door frame between the door and frame. The magnets pull two plates against each other and hold them from sliding across each other. The shear aligning magnetic lock has small pins or dowels that provide a mechanical interference as well as electrical. The shear magnetic lock is sometimes used on glass framed doors or wood doors where the head and plate of the regular electromagnetic lock would not be attractive.

Request to Exit Electromagnetic Lock Delay

In some facilities rather than have the lock immediately release when exit is requested a timer delays the lock release for a few seconds. As described in the courtroom example on the previous page, this has proven effective in some applications. Building code compliance varies from state to state and county to county for this type of installation. Some jurisdictions allow the delay and some do not. Some jurisdictions only allow the delay if there is an audible and visual signal that tells the candidate the door will open after the delay. Again addition of this equipment increases the cost of the electromagnetic lock.

The electromagnetic lock is the simplest of the electric locks. They are strong and they will perform many cycles before needing repair.

ELECTRIC STRIKE

Another method of locking the door and allowing entry is with a device that is called an electric strike (Figure 7-3). The electric strike is a little bit more complicated than the magnetic lock because the electric strike fits into the jamb where the regular strike does. However, the electric strike is usually larger than the mechanical strike which was discussed in Chapter 6, therefore modifications to the door jamb are often required.

The electric strike uses a solenoid to withdraw the strike away from the latch. This releases the door so that it can open. One of the nice things about an electric strike is that if the candidate has a key to unlock the door, the door will continue to operate just as a mechanical lock. But with an electric strike the facility has the advantage of controlling the door for persons who don't have a credential. This type of device is sometimes seen at convenience stores where the attendant has to push a button under the counter to release the strike and the customer can go through. For staff, they usually have their own key and it is not necessary for an attendant to release the door for them.

Like the electromagnetic lock the Facility Manager has to decide how they want the door to perform if power for the electric strike fails. There are two modes for the electric strike, fail secure, which means the lock stays locked (keys and regular hardware will still open the door in this mode). The other mode is: fail unlocked. In this mode if the power fails the door is unlocked because the electric strike is retracted. This door becomes an open door in the fail unlocked mode. Generally, fail secure is the most common mode chosen.

Like magnetic locks, power for the electric strike can be either 12-volt or 24-volt direct current or 120-volt alternating current depending upon the design by the manufacturer. The direct current electric strike is silent while the alternating current electric strike "buzzes" when the strike is extracted. Figure 7-3 is a photograph of an electric strike.

The electric strike also requires wires to provide power for the strike in the jamb. Wires are not required in the door.

The electric strike has a mechanical component and an electrical component and is likely to require more maintenance than the electromagnetic lock. An electric strike can be installed for between $300 and $1000 per door.

ELECTRIFIED BUILDERS' HARDWARE

The third element of electronic/electrical door access control is called electrified builder's hardware. This is the typical door locks/latches and cylinders discussed in the previous chapter, Chapter 6, however in this equipment electrical solenoids and magnets are built into the hardware set. The electrified builder's hardware provides a mechanism to extract the latch or the deadbolt. Electrified builder's hardware fits into the same space

Figure 7-3. Electric strike. (*Photo by Robert Reid, 2004*)

as the mechanical hardware. The cylinder lock, the mortise lock and the deadbolt lock can all be operated electrically instead of with a key. However the problem with these types of locks is that power now has to be run into the door which is more complex than the electromagnetic lock and the electric strike where the power is only run to the frame.

Putting power into the door requires either special hinges that allow embedded wire or, and this is usually the case, an armored loop goes from the frame to the door with enough slack that the door can open and close. This loop exposes a vulnerability to the door, and requires more maintenance than the electromagnetic lock or the electric strike. Plus electrified builder's hardware is more expensive than regular builder's hardware. For some facilities, it can be easier to upgrade the existing mechanical hardware than to install new electrical builder's hardware.

BATTERY OPERATED COMBINATION LOCKS

Some doors have installed self contained hardware that is a battery operated combination lock. The door doesn't require any electrical wiring because the power for the lock is contained in a battery inside the lock. The battery can power the combination or it can also power the combination and the latch. This type of door lock is also discussed more thoroughly in Chapter 8 Electronic Lock Control.

A NOTE ABOUT DOORS

Remember the discussion about fire rating of doors in Chapter 5? Well, modification of a fire rated door can render the door's fire rating invalid. This is the kind of thing building inspectors look for and, if the door is going to be modified for electrical locks, the cuts that have to be made for the wires and hardware can invalidate the fire rating. If a fire rated door is being modified, it is best to contact the door manufacturer to see if they have a modification kit that allows drilling the door without changing its rating.

Figure 7-4. Battery powered combination door lock. (*Photo by Robert Reid, 2004*)

CONCLUSION

Now that we know how the lock works and the tradeoffs of the different types of locks, the next chapter, Electronic Lock Control: Giving Users Access, addresses the how to get the door unlocked for the right persons but keeping it locked from unwanted intruders.

RESOURCES

For additional resources on this topic a number of manufacturer's offer CD ROMs of their catalogs and additional information on their web site.

Builders' Hardware

Schlage Lock Manufacturing Company www.schlagelock.com

Securitron Magnalock Corp. www.securitron.com

Medeco Lock Company www.medeco.com

Kwikset Lock Company www.kwikset.com
Kaba-Ilco Lock Company www.kabailco.com
Electromagnetic Locks
    Doortronic Systems, Inc. www.dortronics.com
    Security Door Controls www.scdecurity.com
    DoorKing, Inc. www.doorking.com
    Securitron Magnalock Corp. www.securitron.com
Electric Strikes
    Rutherford Controls International,
        Corp. www.rutherfordcontrols.com
    Securitron Magnalock Corp. www.securitron.com
Battery Operated Combination Door Locks
    Securitron Magnalock Corp. www.securitron.com

Chapter 8

Electronic Lock Control: Giving Users Access

While a mechanical or electromagnetic lock protects the facility from intruders, it is important to release the lock at the appropriate time. In fact, if the keys or cards do not work properly the facility staff will eventually try to defeat the locking systems because of the frustration experienced when the locks fail to open. For this reason the systems to unlock the door are often more complex than the electrical or electromechanical locks. These control mechanisms include hardware on the inside for egress and emergency egress. On the outside it includes a myriad of devices designed to sense and detect who it is that requests access and whether they have the right to enter.

This chapter takes access control to the next level: from understanding the locking mechanism to understanding the devices that control the lock. Today there are many different types of sensors in use and this technology will become more complex and sophisticated in the future. This chapter covers these devices, starting with the simplest digital combination lock and progressing to the highly complex thumbprint and retinal scanners. The most common devices today are the card reader and the proximity badge.

How well these devices work, how reliable they are and how much they cost is covered in this chapter. This chapter also explains metal detectors since their function is to aid in determining who has authority to enter and who does not.

The devices and their interfaces with the door are addressed in this chapter. The next chapter (Chapter 9: Computers for Access Control) covers integrating the devices with multiple doors and locks for overall site control.

GENERAL TERMINOLOGY
FOR DOOR ACCESS CONTROL

For the purposes of discussion a few general terms are needed to establish a baseline for communication about how these devices work. First, the person that seeks permission to go through a locked door is a candidate. In order to pass through the door the candidate must present a credential. The electronic devices verify the credential and, if it is acceptable, the candidate is allowed to pass. In a very simple way, a person (a candidate) uses his or her key (the credential) and the door hardware recognizes the key (permission granted) and the person turns the key and unlatches the door (candidate passes). All of the devices from the simple key to the most complex bio scan operate on this basic principle.

The simpler access control mechanisms are covered first progressing to the progressively more complex.

KEYPAD

In Chapter 6, Access Hardware, the mechanical keypad or combination lock was discussed. In that example, the devices to unlock the door operated mechanically to release the lock. A person who knew the combination was able to enter it into the lock and turn the handle. The electric version of this device works the same way, only instead of using mechanical linkages and pins; it uses electric signals to do the same thing. The difference with the electric keypad from the mechanical one is that power is required for the electrical one. Sometimes these are provided with a battery and no hard wiring is required. Depending upon the device, batteries are good for several thousand cycles. Eventually, however the batteries have to be changed. These locksets are controlled from one side only; the reverse side opens without the need for an electrical assist. These devices cost approximately $300 per door and the batteries have to be changed routinely. Device manufacturers recommend changing the batteries at least once per year. Figure 8-1 shows a battery operated electronic keypad.

The candidate must know the combination to the lock to

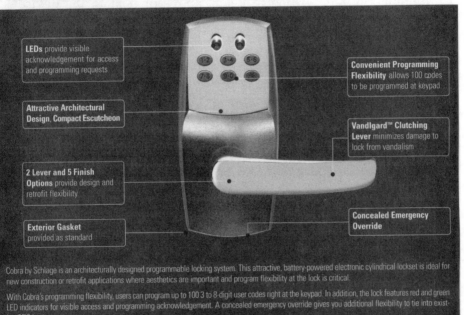

COBRA
BY SCHLAGE®

Attractive to all...but the intruder.

LEDs provide visible acknowledgement for access and programming requests

Convenient Programming Flexibility allows 100 codes to be programmed at keypad

Attractive Architectural Design, Compact Escutcheon

Vandlgard™ Clutching Lever minimizes damage to lock from vandalism

2 Lever and 5 Finish Options provide design and retrofit flexibility

Concealed Emergency Override

Exterior Gasket provided as standard

Cobra by Schlage is an architecturally designed programmable locking system. This attractive, battery-powered electronic cylindrical lockset is ideal for new construction or retrofit applications where aesthetics are important and program flexibility at the lock is critical.

With Cobra's programming flexibility, users can program up to 100 3 to 8-digit user codes right at the keypad. In addition, the lock features red and green LED indicators for visible access and programming acknowledgement. A concealed emergency override gives you additional flexibility to tie into existing SFIC keying systems.

Additional features include extended battery life, Vandlgard™ clutching lever design, a variety of architectural finishes and two lever designs. Cobra by Schlage is an excellent choice for schools, hospitals, multi-family units, retail stores, and commercial and government facilities looking to upgrade their access control from mechanical locks or other manufacturers' locks.

Features	Benefits
• Architectural Design	Flexibility for new or retrofit applications
• Compact cast escutcheon with integrated keypad	Fewer components provide simple and quick installation
• Program up to 100 users at keypad	Minimal restrictions to # of codes and simple, quick programming to add and delete codes
• 2-3 year battery life	Minimal maintenance with standard alkaline batteries
• Vandlgard™ clutching lever	Minimizes damage from vandalism
• LEDs with red/green indicators	Provide visible access and programming acknowledgement
• Concealed emergency override*	Aesthetic appeal
• Adjustable door thickness – 1 1/2" - 2"	Easy installation for a variety of door thicknesses with no extra parts required
• Lever and finish options	Flexibility for new and retrofit applications
• Cover plate options	Masks door preps from other manufacturers' preps while maintaining security

*Requires 6-pin (with spacer) or 7-pin small format interchangeable core for override. Accepts Schlage, Falcon, Best, Arrow or Kaba SFIC cores. Cylinder plug provided standard. Clutch engages lever when cylinder is removed via control key.

Figure 8-1. Battery operated electronic keypad. Reprinted with permission. IR Safety and Security-Americas Copyright 2003. All Rights Reserved

operate the device. However, nothing prevents the candidate from telling someone else the combination. Once the combination is known, anyone who knows it can enter. In extreme cases, the combination gets written in pencil above or next to the device because people forget. The more numbers in the combination, the more difficult it is to remember. Usually the combination is 4 digits but can be as high as 6.

A good security program inspects the vicinity of the doors to make sure the combination isn't posted. If it has been posted, it is necessary to change the combination. On electric devices this is fairly simple. A control code is entered allowing the combination to be changed.

Instead of batteries, the same type of device can be hardwired into the power supply either at or near the door. The voltage and amperage for the device determines the wire sizes. Some power can be fed through special hinges, but usually power for the door lock is fed through a drop from the frame using armored cable.

CARD READERS

Magnetic Card Reader

The next device in order of complexity is a card reader. This is the same device as a credit card with a magnetic stripe on the back. The card is "swiped" through a reader either on the door or next to it and if the credential is approved, the door unlocks. Many office type facilities utilize card readers and make them a part of the person's badge. This way the badge has the person's identity on one side and the magnetic stripe to unlock the door on the other side. The entire process of badging employees is covered in Chapter 15: People Systems. Figure 8-2 shows a magnetic card reader.

The information stored on the card is compared to information stored by the system connected to the reader. If the two items match, then the card is approved and the door unlocks. For an application where only one door is provided with a reader, the confirming information can be stored near the reader in computer memory circuit boards. For an application where one card is read

Figure 8-2. Magnetic Card Reader. This is a photo of the Kaba-Ilco Solitaire 71011 System. Reprinted with permission from Kaba-Ilco Corp. Copyright 2001 Kaba-Ilco Products Catalog, Winston-Salem, NC.

by several readers in different locations, wires or wireless signals can transmit the data to a central location were the information from the card is verified against information stored in a computer data bank. If the information on the card matches the information in the data bank, the door unlocks. The larger the data base the longer it takes for the system to verify the credential but for most systems this is less than two seconds.

Sometimes the card reader is combined with the combination lock. This requires that the person possess both the card and the combination for the system to verify the candidate's credential.

Problems with the magnetic card reader include dirt or elements fouling either the badge or the reader and problems with magnetism of the cards. The card's magnetic memory can be erased or blinded by contact with magnets or by going too close to magnetic alarm devices in grocery and convenience stores. Usually this problem is only temporary and after a few minutes or an hour the card "recovers." But in the meantime, the person's access is going to be denied.

The magnetic card reader does not work well when it is installed outside because it cannot get wet. Dust and bugs also impact its operability outside.

Hotels commonly use the magnetic card reader device for issuing hotel room keys. The lockset is capable of reading the cards. One element of data placed on the hotel card is the time when the card expires. This way when the candidate's stay has ended, the cards no longer grant access. Since most hotel room doors are located inside, this system works fairly well. For hotel rooms with these types of locks in outside service the locks do not perform as well, with many denials of credentials that should be approved. Magnetic card reader locksets sell for $250-$650 depending upon how the cards are programmed.

Wiegand Card and Reader

The Wiegand Card and Reader are similar to other proximity sensor cards. The Wiegand Card uses wires that alter a magnetic field when passed through the field established by the reader. The Wiegand Card and Reader can operate without external power.

RFID or Prox Card and Reader

As result of the problems with the magnetic card reader, the radio frequency identification device (RFID) was developed. The RFID carries information similar to what is carried on the magnetic card. The advantage of the RFID is that it does not have to touch the card reader to have its credential read. The RFID is also called a proximity reader.

The RFID card has a loop of wire sandwiched between layers of plastic in the card that acts as an antenna to receive signals from the reader. The antenna magnifies the signal and replies with data embedded in a micro computer chip in the card. The reader in this case is called a Prox reader because the device doesn't have to actually touch the reader; it only has to come into the proximity of the reader. See Figure 8-3 Proximity Reader.

More data can be stored in the Prox card than in the magnetic card and since the card doesn't have to touch the reader, it is slightly faster than the magnetic card. Since the Prox card does not have to be inserted into the reader, the candidates' hands are free to carry things. The Prox card can also be installed outside with fewer problems than magnetic card readers.

The wiring of the device to read Prox cards is similar to the wiring of the magnetic card reader. The credential of the Prox card can be tied to a combination lock or other second verification that only the candidate knows, otherwise, whoever has the Prox card can obtain access through doors locked and controlled by small Proximity Card readers. In most installations, multiple doors are controlled with an individual Prox card. In an office setting, an individual can access the entire group of perimeter doors using

Figure 8-3. Proximity Reader. This photo shows a typical proximity reader installed in a stairwell of a hospital. The secured area is the maternity ward. (*Photo by Robert Reid, 2004*)

the same card. Another individual with a different card, can also access these same doors. In the case of magnetic cards for a hotel, the magnetic card opens one lockset for the room and perhaps one or two others. The proximity card and reader is capable of many more cycles than the magnetic card reader.

There are two types of proxy cards, passive and active. In the passive Prox card, the information on the card is transmitted via a signal from the reader. For this reason the passive card has to be fairly close to the reader to be read, usually within 5 centimeters. In the case of the active Prox card, a battery in the card sends the credentials to the reader. The active Prox card has a range greater than the passive card and can send signals as much as 500 feet. Active Prox cards are now being used to control facility gates because of their range.

As with the magnetic card, the proximity card can also be an employee badge. The Prox card doesn't have to be taken out of its holder to be presented the way the magnetic card does. Also, some manufacturers offer a key fob that attaches to a person's car keys that acts as a Prox card.

Proximity cards and magnetic cards can be the same size or they can be different sizes. The magnetic cards usually fit into a wallet or billfold. These cards are economical costing from a few cents up to a few dollars and readers cost $125 to $400 per door. Installation is in addition to these figures. More information about cards is presented in Chapter 15: People systems.

Bar Code Reader

For some security applications a bar code reader is used as a credential for access control. The bar code is the same technology used for products in commercial stores for pricing and inventory control. The bar code is a series of black stripes on a white background. In recent years the bar code readers have improved from the sluggish, temperamental readers of a few years ago. The best bar code readers use a laser and a hologram on a spinning disk. This is the technology that enables the bar code to be in multiple positions in convenience and grocery stores. The reader is more complex than what is usually applied as a proxy card or magnetic stripe reader. The bar codes are easily duplicated so they are not often used for access control. Bar code readers cost between $250 and $400.

INTERCOM SYSTEMS

Another way of controlling a door is with an intercom system. This assumes that a receptionist or other staffer is available to respond to the intercom request. A person could recognize the voice of the candidate, or the candidate could present a verbal credential. The receptionist or guard then grants permission to enter the door by releasing an electric strike, magnetic lock or electrified builder's hardware. Intercoms are wired similar to electrical devices. Sometimes it is necessary to shield the intercom from weather. A typical intercom system is encountered in residential gates to subdivisions or commercial gates. The intercom has a call button that works like a doorbell to call the assistant. The problems with the intercom are that there needs to be someone to respond to the call. The advantage is that it is not necessary to issue a magnetic or proxy card to the candidate. The intercom system works best with one or two doors and is not as fast as Prox or magnetic cards because the administrator may not be available to answer the call. Since the candidate does not have to carry a credential, this may be the best choice for the facility.

In the same way that the audible intercom system works, manufacturers are also providing intercoms with video display. With the video display, not only is the voice available as a credential, but, the person to who the credential is presented can visually verify whether the candidate is valid. Intercom systems can be installed for $200 to $600 per door. A typical video intercom is shown in Figure 8-4.

The attended door, where a real live individual attends the

Figure 8-4. Video Intercom Device. The Enterview TI14 Audio Visual Intercom System from Lee Dan Communications, Inc. Reprinted with permission. Copyright 2003 Lee Dan Communications, Inc., New York.

video system for doors and gives permission to enter is discussed more fully in Chapter 11: Controlled Circuit Television Systems.

BIOMETRIC CREDENTIAL VERIFICATION

With the exception of the intercom system which is relatively simple there exist a number of biometric analyzers for facilities seeking higher security than can be provided by keypad or card readers. These bioscan systems require some personal information from the candidate to validate the credential. The personal data can be a palm print or fingerprint from the hand, voice recognition, facial recognition or a retinal or iris scan of the eye. These devices are more complex than the simple systems above and their results are more varied. Generally, these devices take longer to validate the credential than the proximity and magnetic cards and there are more false rejections. In order to speed up the verification process some of these devices request a password or combination. The database then knows whose biometric information to check. How these systems work and the tradeoffs of these systems is explained below.

Palm or Hand Geometry

The palm print or hand scanner reads data from a human hand and compares it to data stored in a database. Information stored includes size and spacing of the fingers. Palm Geometry has been used for many years and has become popular in theme parks for controlling access. The palm reader doesn't recognize the entire hand, just a few key points. Figure 8-5 shows a typical palm reader. The palm reader uses finger pins to set the fingers in position for the reader. Then the palm is scanned three times for 90 characteristics and makes a template that averages the three trials. The hand reader eliminates the need for a prox card or other pass credential. Most of the palm geometry readers check for temperature and pulse to prevent cutting off the hand or using the hand of a dead person. The hand or palm geometry is relatively fast, 3 to 5 seconds, and has a low reject rate. Palm geometry readers cost about $1500 dollars without installation.

Figure 8-5. Palm reader. Reprinted with permission. IR Safety and Security-Americas Copyright 2003. All Rights Reserved

Fingerprint Scanner

For identification and verification one of the oldest and most robust systems for identification is fingerprints. However, legally taking person's fingerprints and placing them into a database has been challenged in some applications as an invasion of privacy. For entry control, biometric scanners can scan one or two fingers and verify the identity of the candidate against a template stored in a database. Fingerprint scanners are reliable but measure more information than Palm Geometry and hence take longer when comparing a credential to a database. Like the palm reader the fingerprint reader requires no badge or password to remember. False acceptance with fingerprint scanners is low but a false rejection sometimes occurs. When this happens the candidate is rejected. Sometimes they can try again immediately and gain access, at other times they are rejected and forced to seek another method of entry. Finger and fingerprint scanners cost $600 to $1000. They are as reliable and robust as the palm geometry readers.

Voice Recognition

Voice recognition systems have been tried and are constantly improving. However, sounds passing through air can be confused by background noises, music, wind on the microphone and other elements. Voice recognition by computers/machines is not recommended for access control or identification.

Retinal Scan

The retina, the pattern of blood vessels on the back of the human eye, is unique to every individual. Access control for doors can be programmed to work with retinal scanner as with palm geometry or fingerprint identity verification systems. The retinal scanner uses infrared light to look past the pupil and iris onto the back of the eyeball. There the imprint of the retina is taken and compared to information in a database. If enough data points match the identity is verified and the door unlocks. The reliability of the system is high with few false positives. But because the scanner must see the retina clearly to take an imprint to compare there are often problems with rejection of legitimate candidates. For this reason the retinal scan can be cumbersome and time consuming.

Some staff may feel the retinal scanner is intrusive as they may believe there is a health threat as a result of the infrared scan of the eye. These devices are safe but a manager may have to deal with employee concerns if this technology is implemented. As a result of these concerns, this type of device is only recommended for very high security applications. Cost of a retinal scanner is about $250 for the camera. Software and installation is extra.

Iris Scan

As with the retinal scan the iris scan examines a unique biometric element and records it for credential verification. The iris of the eye is as unique as the retina or fingerprints and does not change over the individual's lifetime like a facial image does. Infrared light is used to read the iris. Since the iris is near the surface of the eye it is easier for the optical camera to see the iris than the retina.

Problems with the iris scan are the same as problems with the retinal scan, sunglasses or eyeglasses and contact lenses can affect the reading leading to a false rejection. Managers may encounter staff resistance to iris scanning because of fears that the light scanning the iris poses a health hazard to the eyes. Iris scanning is slower and more costly than fingerprint scanning with about equal accuracy.

Facial Recognition

At the present time the Untied States government and other world governments are investing heavily in facial recognition technology in an effort to catch criminals and terrorists. A database of the faces of known or suspected criminals is created. Then faces of crowds are scanned and the faces in the crowd are compared to the database of known faces. If there is a suspected match, a suspect is detained until his identity is confirmed.

This type of security is still developing. Facial recognition is fraught with problems from people wearing sunglasses, or regular glasses to beards, mustaches, contact lenses, changing hairstyles and many other factors. For specific security applications facial recognition is less attractive than the palm or fingerprint reader. The facial recognition system uses a camera to take an image of the face and compares it to data stored in the system. The data stored includes distances between the eyes, length of nose, distance between the nose and lips and many other data points.

The facial recognition system uses much more data than the palm or fingerprint and so it must search longer for the recognition parameters in the database. For facility protection, unless it is frequented by large members of the public, this technology still has a way to go.

HIGHEST LEVEL SECURITY SYSTEMS

Personal Recognition

The highest level of protection for access is personal recognition by armed guards. The chances of being able to fool a security guard who personally knows the candidate is almost zero. However, guards are expensive and keeping the names and faces of staff up to date requires constant training. More information on guards is presented in Chapter 14: Guards and Guard Forces. In other situations, personal recognition by staff for granting access is effective, but untrained or unprepared staff cannot deal with someone who desires entry and is willing to force their way in.

Closed Circuit Television

Highly secure door access control can be obtained from using electric locks of any type coupled with recognition via closed circuit television systems. This application is common in high security applications like jails, prisons and high security control rooms of power and chemical plants. The candidate requests entry by pressing a buzzer and a person from inside the facility looks at the door through a CCTV camera. If the person requesting access is authorized entry, the person releases the door from a central control station. Central control stations and CCTV will be covered more thoroughly in Chapter 11 Closed Circuit Television Systems.

METAL DETECTOR

Finally, many installations use metal detectors at the entrance to prohibit weapons from entering the facility. Metal detectors require guard attendants to monitor and screen the candidates. Figure 8-6 is a photograph of a metal detector station for a court facility. The metal detector is most commonly used to screen the general public in situations where weapons are reasonably expected. The metal detector does not detect non-metallic objects like chemicals or plastic knives. Hence most metal detectors are combined with X-ray machines to scan bags and articles carried by the candidates. The metal detector does not stop the candidate the way a locked door does. The metal detector relies upon the guards to halt the candidate, if they are suspected of carrying weapons.

While some metal detectors are rated for service outdoors, most are designed for use indoors. The metal detector uses coils to establish a magnetic field through which the candidate passes. A metal object passing through the magnetic field disturbs the field. The disturbed field signals an alarm. The coils are set in several zones, the head, the chest, the waist, the legs and the feet. The disturbance of the coil signals lights to flash on the back side

Figure 8-6. Metal detector station. (*Photo by Robert Reid, 2004*)

of the detector where an attendant can see where the metal object is. This allows the attendant to concentrate on the location where the metal object is suspected. Each coil can be set differently to measure small, medium or large metallic objects.

Some metal detectors can have the bottom field disturbed by liquids from mopping the floor, so mopping near the base of a metal detector should be performed with care. Metal detectors can also be programmed to count the number of people passing through and provide data on the number of people stopped for presence of metal objects. An expensive metal detector costs $7,500 dollars. An inexpensive one with fewer coils and less sensitivity can be purchased for $1,800.

OTHER DEVICES TO SCREEN AND
PROTECT THE DOOR FROM INTRUDERS

New research is being conducted on access control continuously. It is possible to combine two or more of these devices. Using the keypad with the thumbprint reader for example, requires the candidate must possess both credentials to pass. More sophisticated devices are being developed.

The US State Department and the US Military are developing newer sophisticated equipment and this area of physical security is expected to change rapidly over the next few years.

Throughout this chapter data has been presented on how a candidate presents his credentials for access. However, if there are more than a few doors and a number of devices are installed, signals must pass between the devices and the doors. The next chapter addresses wiring and computer control of doors.

RESOURCES

The Biometrics Catalog http://www.biometricscatalog.org/is a U.S. Government sponsored database of information about biometric technologies including research and evaluation reports, government documents, legislative text, news articles, conference presentations and vendors/consultants.

The Department of Defense has a biometrics web site where reports on biometric testing are stored. http://www.biometrics.dod.mil/

One of the Proximity Card vendors is Data Card Systems, Inc. www.datacard.com

Chapter 9
Computerized Access Control

Now that tools for credential verification are known, the next step for the security professional is to centralize the system. Since so much of the information relative to individual credentials and access is needed, a database program is used to store the information.

At the end of Chapter 6 a key control program was discussed. With the electronic door locks and readers it is still necessary to collect the same information. Who has access to the facility? (Who has the key?) Who granted the access? When is access authorized?

With electronic lock control systems like magnetic cards, proximity cards and biometric scanners, it becomes possible not only to control who has access but the time of day when access is granted. Locks can be set for weekday hours only for example.

DATABASE OPTIONS

As with keys, a database is used to store information about employees' access and this can be used to control the door locks. All of the information needed for key control fits neatly into the database, as well as, credentials present in cards or resulting from the bioscan templates. Then when a credential is presented the database is searched.

If the credentials are valid, and it is the right time of day for that door, the database sends a signal and the lock on the door is released.

COMPUTERS

One of the first decisions required of the facility manager for computerized access control is what computers to use. Most of the commercial database/door control programs run on standard personal computers. There are two ways for the facility to go with use of computers to control the doors. First, should the computer be a stand alone computer dedicated solely to the security mission or should the computer be usable by administrative or security staff for other chores? The computers can certainly handle multiple tasking, but should they?

Most security professionals make a request that the computer system be a stand alone system dedicated to the security mission. Personnel data related to access control is stored in the security system and this system is dedicated solely to the security mission. In other facilities, the security software is one of many programs on the system and the computer is used for other tasks like email, word processing and internet. Ultimately the decision is up to the facility. The decision is a function of the number of staff, whether the staff is fully dedicated to security, or whether the doors are used by members of the public. Using the risk assessment tools developed as shown in Chapter 1 helps the facility make these decisions. With risk assessment information, the decision process is easier.

In addition to the decision of whether to use stand alone computers, some facilities decide that they need proprietary software to contain the data and control the doors. The computer program is written "in house" for that facility only and uses its own generated computer code. This is, of course, much more expensive than commercial software that is publicly available but since it is stand alone, it would be much more secure. Commercial software programs are constantly being updated and software that is 10 years old will not run on new computers today. This factor should also be taken into consideration when evaluating software for door control.

The software is also capable of many other tasks.

Recording
In addition to scanning the credential to determine the

candidate's validity, the software can be set to record the name and time of entry. Some proxy readers are used for timekeeping and payroll activity. Also the system can be set to provide higher levels of access to managers. New people can be allowed access during normal business hours while supervisors and managers can have access earlier in the day and later in the evening and on weekends. Finally, janitorial staff can have access outside normal business hours for cleaning on nights and weekends. The software can record who requested access to the door and what time the access was granted. It can also be set to record who requested access and when access was denied. Finally staff movement patterns can be analyzed to improve security.

For a large facility, with many access points and several levels of security, it may be preferable to provide separate systems that operate autonomously and independently. The more complex the information and the more doors and access control parameters the software has to search, the longer it can take for the computers to validate the credentials.

At one facility the recordings of door openings and closings were tied to work permits to confirm two employees that had violated access control procedures. Since the violation was serious, the employees were given the option to resign or have their employment terminated. Given that the evidence clearly showed the doors opening and closing when the adverse events took place and that they had the work permit to unlock and open the doors, the employees chose to resign and learned a valuable lesson.

Timers

In addition to recording the opening and closing of doors and by whom the doors were accessed, the software can set timers to keep track of how long the person has been inside, and timers can be set on the doors themselves. If the door remains open too long, a signal can be generated that says the door has been propped open or has failed to close properly. Maintenance or security can then be dispatched to confirm the door has not been inadvertently left open.

Interlocks

An interlock is a recording of two events that are related with a pause or a hold point that one event must finish before a second event can begin. In a chemical processing building or clean room for example, an interlock can be installed that prevents both doors of an airlock from being opened at the same time. The interlock sees that the first door is open, and even through it receives a valid credential to open the second door, the interlock maintains the lock on one door until the second door closes. In another scenario, it may be necessary to close both air lock doors for a period of time necessary for the ventilation system to completely change the air in the room before opening one of the doors. A timer set for the amount of time necessary to sweep the room keeps the doors locked until the air in the room has been completely swept out by the ventilation system.

Other interlocks can be tied to other building systems like the fire alarm or the air conditioning systems; i.e. the system could be programmed to unlock all the doors if there is a fire alarm.

Badging

Software vendors that sell door control systems usually offer an option to generate the badges. The data about the person kept in the door control database can be generated onto the badge which is then used as the proxy card or magnetic card. Badging options can include photographs. The decision to badge employees, what data belong on the badges and how much badging costs is addressed later in Chapter 15: People Systems.

Mapping

Some door control software contains floor plans or maps of the building showing which doors are opening or closing. The software can be tied to an alarm system showing what area is in alarm and what the alarm is.

Mustering

One of the options available on computer systems that track who has entered and exited is called a mustering option. The software tracks the names and number of people in a designated space. If it is necessary to evacuate, the mustering option can prepare a list of individuals in order for the facility to verify everyone has been evacuated. Chapter 16: Emergency Response talks about verification of evacuation in the emergency response planning process.

WIRING FOR DOOR SIGNALING SYSTEMS

In order to carry the signals from the readers to the doors and from the doors back to the locks, wiring or signal cabling is necessary. The information can be carried with electrical power wire, coaxial cable, fiber optic cable (although this does not carry the power to energize locks) and wireless radio frequency energy.

Wiring

For the signals for door locks and credential transmission, wires are used to carry the information from the readers at the doors to the computers and from the computers back to the locks. Most of this wiring is low voltage (24 volts direct current or 24v DC) where normal wiring in a building is usually 120 volts alternating current (120v AC). Transformers are used to lower or step down the AC voltage to either 12 or 24 volts DC.

Because signaling systems are low voltage, the wire type is also different, usually smaller, lighter and easier to install. The wire is usually shielded to prevent signal noise from interfering with the communication between the computers and the doors. Shielding is foil wrapped around the wires to protect them from stray signals.

Depending upon the distance to transmit the signal it may be necessary to provide in-line amplification to enhance the signals.

Unshielded Twisted Pair Cable

Some devices use what is called unshielded twisted pairs. This wire is typical of wire used for telephones, and some facilities have used spare telephone cables as supplemental wiring instead of installing new wires. Figure 9-1 is typical of unshielded twisted pairs.

Coaxial Cable

Coaxial cable, commonly used for television signals, can be used for some door and reader signals. Coaxial cable or Coax, as

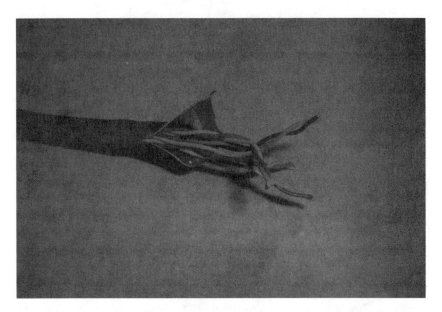

Figure 9-1. Unshielded Twisted Pair Wire. This photo of what is called CAT 5 cable contains four pairs of wires. *(Photo by Robert Reid, 2004)*

it is often called by technicians, is a single conductor, usually encased in plastic, surrounded by a wire mesh shield. This shield protects the conductor cable from electromagnetic interference. Problems with coaxial cable include failure where connectors are attached so maintenance needs to be performed on the cable connectors periodically.

> Coaxial cable can sometimes fail at the crimp station where the cable terminates. At a facility with highly complex control system this problem was finally diagnosed as a cable connector failure. The cable was a loop with over 200 connectors that were daisy chained through many devices. Maintenance forces had a tool to check the cable for good signal but it was very tedious to check each connector. The loop was broken into segments and checked piece by piece. The problem existed for several weeks but only after detailed review was the problem correctly diagnosed. Others in the organization who could not solve the problem had determined it was a system problem that could not be solved. After a complete rundown of the system the bad connectors were isolated and replaced. Figure 9-2 shows Coax Cable Connectors.

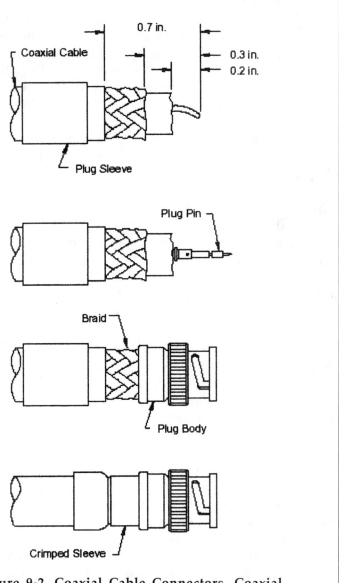

Figure 9-2. Coaxial Cable Connectors. Coaxial Cable is another type of wire, used as data highway but also for Closed Circuit Television Signals. Transmitted with personal correspondence. 2004.

Fiber Optic Cable

Fiber optic cable is used for data communication between central stations and control panels. The biggest advantage is that fiber optic cable is difficult to tap and therefore it is more secure. The cable is self monitoring, and not subject to radio frequency interference like wire. However, many of the solenoids and devices that control doors cannot react to a signal from a fiber optic cable. The electromagnetic lock, for instance, cannot be operated without an electrical power supply. However, the door controller can be signaled from fiber optic cable and local power can be triggered to release the lock.

Wireless

Newer devices have been able to utilize wireless technology to transmit the credential and door control signals without wire. Wireless uses radio waves and is encrypted between the transmitter and receiver so that the signal is protected. However, wireless signals are subject to interference. A major function of the wireless technology is the application of the antenna. The antenna has to be sized correctly to receive the correct signal and there are several types of antennas. The larger antennas are used for the longer range. Wireless is also the technology used for cellular telephones and computer WiFi applications. Right now the use of wireless signaling devices is one of the most dynamic and fast growing elements of communication technology.

CONTROL STATIONS

When multiple door controls and readers are used, a control station is needed. The control station can be at a guard station, the desk of the loss prevention officer or at an administrative person's desk depending upon the access control complexity. The control station is essentially a personal computer workstation with access control software loaded on it. Controllers at the doors are connected to the PC to allow communications between the computer and the doors. For more complex systems some computers can interface with the intrusion detection devices and systems explained in Chapter 10: Intrusion Detection.

For many facilities, computerized access control is tied to the employee badging process. A complete explanation of employee badging is included in Chapter 15: People Systems. For many facilities employee badging is an administrative function performed by administrative or human relations staff as an ancillary duty after the policy for access control has been established. The control station is usually the computer that responds to requests from the door readers with credentials. The computer checks the data from the reader and if the credential is valid, the computer sends the signal to the door, releasing it. Usually, if the credential is not valid, the door remains locked but some software programs record who presented the credential and the time of day.

There are many vendors of door control software with various types of computer code architecture. Some are dedicated to read only devices from a few manufacturers; others are designed to accept signals from multiple types of devices. The software can be specifically written for that facility or it can be a commercial product from one of the vendors. For door control, the software usually runs on a personal computer that is operating one of the commercial software applications such as Microsoft™, UNIX™, or the popular Linux™ based operating system.

The control station also receives the data for door credential verification and accepts entry of new codes for new employees with new credentials. Some facilities limit the acceptance of new credentials to one or two days per week, rather than staffing the facility full time.

Usually the access control software is running an internal clock and can lock all the doors after business hours, releasing them again in the morning. Some applications can have settings allowing some personnel access 24 hours per day, while others, require special permission to enter the building outside normal business hours.

For control stations combined with a guard force or intrusion detection systems, additional hardware and software is provided including closed circuit television monitors.

For facilities that need an extra layer of protection, an intrusion detection system is installed that uses sensors to detect intruders outside of normal business hours. The next chapter addresses intrusion detection systems more fully.

Figure 9-3. Control room station. (*Photo by Robert Reid, 2004***)**

Control room alarm software screen colors are chosen to indicate to operators by color the seriousness of an alarm. If all is satisfactory and stable, the color blue is used. For a situation that is changing but it is acceptable to change, the color green is used. For a warning, the color yellow is used indicating a possible problem. For an out of specification operation, the color red is used. Usually for control stations a condition flashes until the operator acknowledges the situation, then the color stays solid until it clears. Some software requires the color to remain until physically cleared by the control room operator. All warnings and alarms are recorded along with the keystrokes of the operator and times.

Chapter 10

Intrusion Detection: Preventing Unauthorized Access

INTRUSION DETECTION SYSTEMS

In order for a facility to protect assets, the next step to prevent an intruder is a system that detects unauthorized access and notifies the authorities. In the previous chapters, methods to lock the doors and allow only persons with properly authorized credentials to pass were examined. This chapter takes the next step in physical security by explaining the tools and techniques available to detect unauthorized access and notify the authority. Today, as in the past, security guards continue to patrol facilities looking for intruders, but many new electronic devices make the security guard's work easier and in some cases these new systems have been able to eliminate the need for security patrols entirely.

The electronic intrusion detection system is a combination of devices to detect an intruder, a signal system to carry the information from the sensor to a central processor or device, and a method of notifying security forces or local police that an intruder has attempted or has accessed the property. The total design of the system depends upon the response time of assistance. No IDS system is effective if no one responds.

RESPONSE TIME

Recall that the risk assessment in Chapter 1 included a review of local authority or guard force ability to respond to an event. An intrusion detection system must be capable of detecting an intruder and dispatching a response team within the time required to apprehend or stop the intruders. In many applications

In a highly publicized crime in Oregon, intruders entered a facility that appeared to be protected with a surveillance camera. In the course of robbing the facility, they were disturbed by the owner. The intruders shot the owner and left the facility, retreating to a bluff overlooking the location to determine if the police were called by the IDS system. After an hour, when no help arrived, the criminals returned to the facility and continued to burglarize the property, while the owner lay dying from his wounds. Had the owner connected the device to a notification system, the police would have arrived in time to rescue the owner and apprehend the criminals.

the IDS system notifies authorities with the objective to prevent the intruder from exiting with the asset rather than preventing the intruder from gaining access to the asset. The response time is the total time for the authority to respond to the alarms. The IDS system must include the steps of detecting intrusion, validating the information and notifying the authority. Then the facility must wait for the response. Even though response times may be very short, i.e. less than 10 minutes, in that time an intruder can scale an 8 foot fence and run a mile. With an automobile the intruder can be even farther away. This is the reason fences, gates, door and window locks are important elements to delay the intruder.

However, without response and with unlimited time, an intruder can eventually gain the asset. An intrusion detection system is one of the methods for signaling for aid. Some IDS devices can take automatic action such as turning on flashing lights or sounding horns to call attention to the facility. These actions can act as deterrents for some intruders. But for an intrusion detection system to take actions designed to harm someone is against the law. Booby traps are not allowed. Booby traps include trap doors, guns wired to door latches, weights, blades or other devices that could harm a person. This is because there may be a legitimate excuse for the intrusion. The employee returning because he forgot to turn off the lights should not be harmed.

Knowing the response time is an important element of IDS system design. Most facilities design the IDS from the inside out. The system consists of successive perimeter rings each further out from the former until a series of concentric rings surround the asset. The first part of the IDS system is the sensors that detect the intrusion.

SENSORS

Since the IDS system is a series of concentric rings radiating out from the asset, the IDS devices each perform different functions depending upon where they are in the ring. For lower technology systems, one or two devices can be used.

IDS sensors work via several methods depending upon what needs to be detected. Also some IDS sensors are designed for indoor service, others for outdoor service. The devices designed for outdoor service are more expensive than indoor devices because the outdoor devices must work in all types of weather. Depending upon the device the outdoor sensors may give false alarms because of wind, dust, snow, blowing leaves or rain.

Types of sensors include door and window contacts, passive infrared detectors, microwave detectors, sound or ultrasound detectors and grid sensors. Perimeter detectors include fence sensors to detect intruders at the fence, microwave and beam detectors to detect persons in the perimeter. Door contacts can be used on gates as well as doors.

Door and Window Contacts

For a facility with door and window locks, the simplest intrusion detection devices are electrical contacts on the doors or windows. If the circuit is on, the electrical loop is broken when the door or window with electrical contacts is opened. Many of the magnetic locks discussed in the previous chapter have a door contact switch included. So the magnetic lock performs as a lock and also doubles as a door contact for the IDS system.

Balanced Magnetic Switches

For a more sophisticated application a door can have a device called a balanced magnetic switch installed. This device has a switch held in position with a magnet on the door or window. When the door is opened, the magnet can't hold the switch and the contact closes indicating the door has been opened. The advantage of the balanced magnetic switch is that it does not require wires in the movable part of the opening the way electrical contacts do. The wiring only needs to be fed to the frame.

Passive Infrared Detectors

One of the most popular detectors for inside service and also applicable for outdoor service is the passive infrared detector, also referred to as a PIR detector. Various models of PIR sensors read at different distances and widths. Some devices are capable of having the zones adjusted in the field. The passive infrared detector reads infrared energy in the zone viewed by the device. A live animal emits heat in a specific thermal range for which the device is tuned. PIR Detectors are passive because the sensor reads thermal energy in the environment, it does not send out a signal the way a microwave transmitter does. Newer PIR detectors are preprogrammed for pet immunity. This means the device is set for the thermal radiation given off by humans. The signal transmitted by pets is in a different band. PIRs can have difficulty in cold weather, snow, fog or rain. The PIR can be triggered by radio frequency interference so some installations require a signal stabilizer to reduce the interference. PIR detectors for a 30' by 30' zone cost between $20 and $50 dollars. Long range PIR detectors, up to 300' cost $175. Figure 10-1 shows the field pattern of one type of passive infrared sensing device. Passive infrared devices are also used for outside service. Some commercial devices come with a light and a passive infrared sensor. Their function simply turns on a light to illuminate the area when someone enters the sensor field. While this device won't deter a determined intruder it may cause the more timid intruder to withdraw. Figure 10-2 shows a typical Passive Infrared Sensor.

Microwave Motion Detector

The microwave motion detector is an active transmission device that sends a signal out into a zone and monitors that zone for disturbances in the field. The field can be tuned to be narrow or wide, set for near or far sensing. Internal microwave motion detectors are passive sensing, reading the reflecting signal from the transmitter and reading changes in the field strength caused by an intruder. For outside service either passive or active micro-

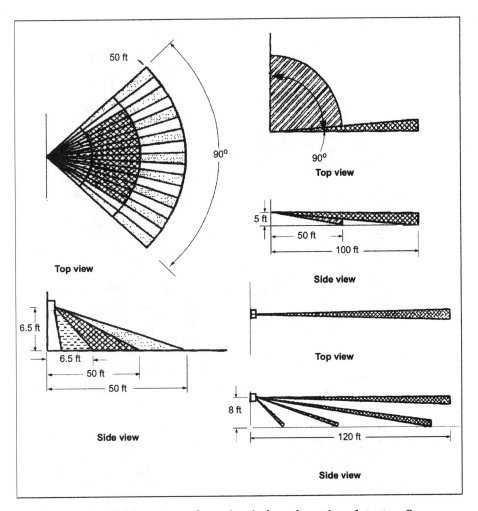

Figure 10-1. Field pattern of passive infrared motion detector. Source: US Army Field Manual FM 3-19.30.

wave sensing can be used. Active sensing uses a receiver at the other end of the range being monitored. When an intruder enters the field and disturbs it, the disturbance is read at the receiver. Some organizations use microwave transmitters near fences because of the long straight run. Some microwave sensors are combined with passive infrared motion detectors to reduce false alarm signals. A combined PIR/Microwave detector for inside use

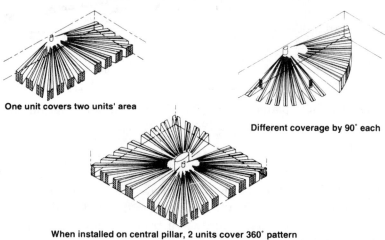

One unit covers two units' area

Different coverage by 90˚ each

When installed on central pillar, 2 units cover 360˚ pattern

Figure 10-2. Passive infrared sensor. This is a Takex America, Inc., passive infrared sensor model PA4701.

sells for about the same price as the PIR detector. Outside active service microwave transmitters and receivers are more expensive. Stray radio frequency emissions can disturb the field of a microwave sensor so the field has to be tuned by trained personnel.

Photoelectric Beam Sensors

Photoelectric beam sensors use light or infrared beams to establish a pattern between a sending unit and a receiving unit. If an object breaks the beam, an alarm is triggered. Beam sensors are usually installed near gates or doors as a backup to the physical gate. Figure 10-3 shows a typical type of beam pattern for a high security gate. On a smaller scale, this type of device is fairly typical of the garage door object sensor. When a person breaks the beam of the garage door photoelectric beam sensor, the door stops closing to prevent injury to a pet or small child.

Ultrasound Detectors

In addition to the light sensors, an alarm system can include sound sensors that detect the noise of breaking glass, drilling, digging, or cutting. Ultrasound detectors can be installed within the building's walls or in rooms. Some ultrasound devices are installed underground and detect digging or burrowing.

Glassbreak Detectors

The glassbreak detector is a specifically tuned ultrasound detector designed to listen for the sound of breaking glass. When glass is broken the detector's contacts close signaling an alarm. Glassbreak detectors can include two microphones to read the sound and compare them for a reduction in false alarms. Glassbreak detectors are usually installed inside. These devices sell for $30 to $50 dollars.

Capacitance Detectors

Another type of IDS sensor is the capacitance detector. This device is designed to detect an object being touched or handled. For example a file cabinet can have a capacitance detector set to read a small electrical charge on the cabinet. When the cabinet is touched, the intruder grounds the cabinet and the capacitance sensor triggers an alarm. Capacitance sensors need for the object

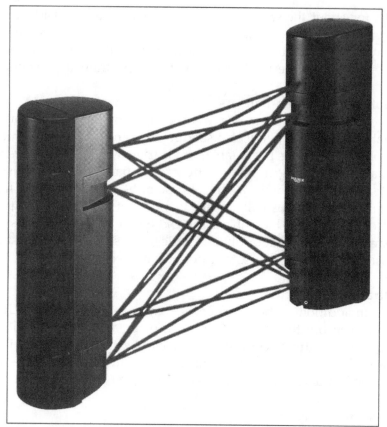

Patent registered: 1, Utility model registered: 7, Utility model pending: 6,
Design pending: 1.

Figure 10-3. Field pattern of photoelectric bean sensor. Source: Takex America, Inc., quad-synchronized photoelectric beam sensor company brochure, Copyright 2003 Takex America, Inc., Sunnyvale, CA.

to be isolated to work properly, hence the device has to be electrically isolated by setting it on an insulated mat or bumpers. Changes in humidity can affect the conductivity of electrical energy in the vicinity of the device being monitored.

The capacitance sensor can also be used as a fence or boundary detector by setting sense wires as outriggers on fences, walls or rooftops. A charged wire, low voltage, establishes a radio-frequency field between the charge wire and the sense wire. An intruder between the charge wire and the sense wire will interrupt the radio-frequency between the wires triggering a detection signal. This system can be upset by adverse weather, wind, rain or snow so as validation, two loops are commonly installed and global changes to each loop are compared to each other. Since ambient conditions change at about the same rate, the system is set such that when the fields change in the same portion to each other an alarm is not sent. However, if there is a change in signal on one loop that is not matched by the other loop the processor triggers the alarm.

Video Motion Sensors

Closed circuit television cameras (CCTV) can be set to detect motion in the camera's field of vision. The device measures the pixel count and a change in the pixel count over time indicates motion. If enough pixels change, as a result of movement, the device sends a signal to the controller. Since CCTV is such a broad field, a separate chapter is included to address CCTV Systems. See Chapter 11: Closed Circuit Television systems.

TAMPER PROTECTION

Since the sensors advise when there is an intruder, it is only natural for intruders contemplating a break in to try to defeat the sensor. For this reason most IDS sensing devices should be provided with what is called tamper protection. Tamper protection consists of switches or devices that determine whether someone has tampered with the device to render it useless. Tamper protection can consist of case sensors, essentially additional IDS switches; however, many modern IDS devices include internal status switching. The internal status switching tells the central processor that the device is working correctly. If the device is not working correctly, a signal is generated indicating improper operation.

DURESS ALARMS

The duress alarm is a device used to send an emergency signal to a guard force, management or the police that an intruder is present and is acting as a threat. A duress alarm is commonly installed in banks or control rooms where an intruder who could perform bodily harm to the occupants exists. The duress alarm protects not only from an intruder but is used in the event the threat is from a person with valid credentials. For example, an extremely upset individual who commonly works in the facility would have a valid credential, even with the most sophisticated biometric device. As a result the duress alarm is a method for the occupants to notify the authorities that people in the facility are under duress.

The duress alarm is usually a simple button or switch than can be activated by the hand or foot. Usually the duress alarm is under the counter out of sight. It is recommended that the duress alarm be foot activated since and intruder could direct a person to keep their hands in plain sight, making it impossible to trigger the alarm.

A wireless version of the duress alarm is now available for executives, businessmen, couriers and individuals who are facing high risk. The wireless device is a button on a key fob or a device that looks like a garage door opener. The device is hand carried. Pressing a button on the device sends a silent signal to the guard force or bodyguards that rescue is necessary.

Duress alarms should not trigger an audible or visible alarm in the vicinity of the intruder as the sudden ringing of bells or sirens could startle the assailant into violence.

FENCE SENSORS

For high security facilities, fence sensors are sometimes installed to provide an indication of a possible intruder. Some facilities may choose to have a double row of fences with the fence sensors on the inner fence to eliminate false alarms from people casually passing by and simply touching the fence. In remote locations fence sensors can signal an intruder but they can also be fooled into a false alarm by a large animal touching or leaning against the fence.

Taught Wire Fence Sensor

The taught wire fence sensor is a cable stretched between

posts that has a strain gauge on it. If the fence is touched, as an intruder would do climbing the fence, the strain gauge sends a signal to the control device.

Strain Sensitive Cable

The strain sensitive cable is a specially manufactured cable that has set characteristics of electrical resistance and voltage under strain. These parameters change if the cable is stretched, pulled or moved beyond limits.

Fiber Optic Cable Sensor

The fiber optic cable sensor operates similarly to the strain sensitive cable in that the light passing through the cable is subject to fluctuation when the cable is pulled or pushed as it would be if the fence were climbed or cut. The advantage of the fiber optic cable sensor is that it uses modulated light instead of electrical signals and as such it is not subject to interference from stray electrical or radio frequency signals or lightning.

DATA TRANSMISSION

Once the sensor has detected an intruder, the next step is to transmit this information. For sensors, once the device has triggered, the sensor has done its job. In the case of window contacts, the continuous electrical loop is broken, for passive infrared motion detectors; once the movement has been detected the device opens or closes a circuit. This signal is being read at the other end and another electronic device takes the next step. Today, this usually means that a computer receives the signal and begins to process the information. Since a computer can be programmed to perform many functions, the next steps are the result of what has been programmed into the computer's memory. For most facilities the computer's program is a commercial package designed to work with the system to meet the owner's needs or wants.

For a small system with few devices, a computer may not be necessary. The wiring can be set up to turn on bells, whistles or lights. A computer controlled system, however, can be programmed to dial a telephone number and give a voice mail message, page someone via a pager, or send an email and turn on lights, bells or sirens. The computer can take over and transmit signals to lock or unlock doors and perform other tasks depending upon the desires of the facility and the flexibility of the program.

For highly secure facilities, guards are notified and they begin to respond according to protocols required by the designer. Some of the more typical responses and follow up action follow. However, each facility should determine for itself, based upon the risk assessment and the system design, what needs to be done when the IDS system detectors go into alarm.

Location of Alarm

For large systems with many detectors, the first thing that has to happen is that the system must figure out where the alarm has occurred. On a large facility with many openings, or a large perimeter boundary, it does not do much good to signal that a door has been opened or a fence has been climbed when there are many doors or miles of fence. So the first step that has to be performed is a determination of where the intruder is from an analysis of the alarm.

For most modern systems the device has an electronic address in it. This address is a signature that says, "I am the device at the back door and I just saw some movement." The commercial computer program interprets this signal and displays it for human interpretation.

Mapping

Most of the modern commercial computer packages now have a mapping option within them. What the computer does is display a floor plan of the facility and a flashing red light indicates where the alarm occurred. This graphical interface is easier for the guard or attendant force to review than a printout of the location of the device. The computer records the times and locations of devices that have gone into alarm.

AUTOMATIC COMMUNICATION.

The facility manager has several options when it comes to notification. First of all the computer can be programmed to take some actions in response to the alarm. If there is a guard force, the alarm on the computer panel is enough. The guard, watchman or night attendant can respond to the computer by acknowledging the alarm. For some systems the alarm flashes red until acknowledged and after the console operator acknowledges the alarm, the device stays red but stops flashing. While guards and patrols are discussed in Chapter 15, People Systems, in this case, a guard can send a response team (other guards) to the location.

For small systems that can't afford their own guard force, commercial security contractors can provide alarm system monitoring for a nominal fee. Usually this cost is between $25 and $100 per month. The contractor has an office that provides round the clock monitoring. The facility's computer system sends the alarm signals to the contractor's office where manned personnel receive the alarm. Usually the monitoring services will notify the police who will dispatch an officer to the location of the alarm to check out the facility.

In addition the computer can be programmed to call a list of people in order. The list may include the chief of security, the company president, vice president or other party. The computer can be programmed to call and keep calling until one or more people have been contacted.

Problem with Automatic Calling

If a facility decides to have the computer control system call for assistance, this can block the ability of people in the building to call for help if there is an intruder. So if the option of calling is selected, there must be a means of overriding it if a higher priority call is needed.

For the action of apprehending criminals, that is a matter for law enforcement. More about the actions of people is addressed in Chapter 14: Guards and Guard Forces and Chapter 15: People Systems.

Latching and Nonlatching Alarms

One of the problems with alarm devices is latching the alarm. When a device goes into alarm it needs to remain in alarm until the alarm is cleared. If motion is detected, for example, and the intruder stops moving the alarm will clear causing the guards to think the alarm was a system anomaly and not an actual alarm. Another example is with the simple door contacts. The device will alarm when the door is opened, but if the electronic signal does not latch, the alarm will clear again after the door closes and the contacts touch again.

If the alarm device is a latching device, it will remain in alarm until cleared. Usually the alarm is cleared from the control panel since it is easier to clear from the control panel than to physically climb up to where the device is located and reset it.

False Alarm

The false alarm is one of the most troublesome of signals from the electronic intrusion detection system. From the previous discussion on devices, each type of device can be triggered into signaling an alarm when there is no intruder. Motion detectors are susceptible to movement of water rippling for example. In this case there is movement, but the movement is benign. In order to reduce the number of false alarms the passive infrared detector is often combined with a microwave motion sensor. The probability of an event not being an intruder triggering both devices is lower. With sophistication of systems advancing, the number of false alarms has gone down but there are still far more false alarms than real ones. One of the major problems with false alarms is that a system with many false alarms becomes as unreliable as a facility without a system. False alarms should be treated seriously but many facilities start having false alarms on the weekend and no one from maintenance is available to repair the device until the following work week. The cost to troubleshoot false alarms is expensive.

For most law enforcement agencies and some private guard forces a limit is placed upon the number of false alarms to which the force will respond to as a courtesy. Beyond that number there is a charge for responding to a false alarm.

Supervision of Circuits

Intrusion detection devices can be electrically supervised similar to supervision of circuits in a fire alarm system. However, it is not a code requirement to supervise security circuits. Since many of the detectors in system are addressable, the system can

---

A Typical False Alarm Report
Notice of False Alarm

City of _____ State of _____ Security Department
Business or Resident Name _____ Case # _____
Location _____ Date _____ Time _____
Reporting Officer _____ Badge No. _____
Form given to: Responsible person _____ Left on Door _____
Location Type: Business _____ Residence _____ Alarm Co. ____
Weather Conditions: Clear___ Rain ___ Snow ___ Windy ___Thunder ___
There was a false alarm at this location on the date and time shown.
Excessive false alarms are subject to fines. For false alarms in excess of 6
in a 12-month period, security will cease to respond to the alarms until
proper alarm notification can be demonstrated.
Please call the following number if you have any questions: 334-XXXX.

---

call the device electrically and confirm the device is working. This is usually performed as an internal system check and does not require any operator intervention.

Similar to the wiring of a fire alarm system, there are two types of loops. The most secure is the loop that ties back to the main panel. The full loop allows the system to look at the detector from either direction and if there is a break in the loop the device can report via the alternate loop.

A second less costly system is to install an end-of-line resistor after the last device in the loop. The system measures the re-

---

While working in a plant in Utah in the early 1990s the operations staff used walkie talkie radios to communicate while troubleshooting operational equipment. Sometimes the walkie talkie radios would trigger the fire alarms, resulting in the closing of fire dampers, window shutters, and shut down the air handling equipment. Eventually the walkie talkie radios had to be abandoned and a site telephone system used instead for communications to prevent false alarms. Two way radio communications has to be checked in the vicinity of detectors to verify that the signal transmitted from a radio does not trigger a false alarm.

sistance of the loop always expecting to see the resistance of the end-of-line device. If the line breaks or a device fails, the devices between the break and the controller continue to report correctly but devices beyond the break in the loop will not report correctly.

Trouble Alarm

If the loop is broken or one of the devices fails to report correctly, the system will report a trouble signal. The trouble signal is similar to the trouble signal experienced in fire alarm systems. A trouble signal indicates there is a problem with the device or the loop. The device may report correctly, or it may not. In some systems when a device malfunctions, the device can be removed from the system and the rest of the system continues to operate. In this case the system will indicate trouble but not alarm. With the device removed, the loop is still intact so that alarms by other devices on the system still report. In addition, devices will have internal tamper protection, such that, if someone tampers with the device, it will report trouble. A device with a trouble signal may or may not report correctly so it is not an alarm: it is just the system diagnosing itself.

SWITCHING FROM ACCESS TO SECURE MODE

For the IDS system to work correctly, it must be initialized correctly. IDS is usually turned off during the day, and is utilized only on nights and weekends. Some portions may need to be on only at night. Usually the vendor software enables designated users to set the hours when the IDS will be on and when the IDS is disabled.

CONCLUSION

The intrusion detection system can be a valuable asset to the facility. It can warn of intruders and provide time for a response to prevent loss of important assets. For facilities that need high security it can save facility money by reducing or in some cases eliminating the guard force. The IDS can work well with a guard force making their job safer and more effective. But for some facilities theft is possible during business hours and it isn't possible to arm an IDS system because out of the next hundred visitors only one might intend to compromise the assets. For this reason something different is required. The closed circuit television system, while it sometimes acts as an IDS motion detector device, is the next major step in facility protection.

RESOURCES

IDS Alarm System Vendors:
ADT Security Services Corporation www.adt.com
Tyco Fire and Security Systems www.tycofireandsecurity.com
Bosch Security Systems www.boschesecurity.com
Takex America, Inc www.takex.com

John E. Traister & Terry Kennedy *Low Voltage Wiring, 3rd Edition,* McGraw-Hill, NY, 2003
The National Electrical Code® NFPA 70 published by the National Fire Protection Association, Batterymarch Park, MA 2004.
Physical Security US Army Field Manual FM 3-19.30 HQ TRADOC. Commandant, US Army Military Police School (USAMPS), ATTN: ATSJMP-TD, Directorate of Training, 401 Engineer Loop, Suite 2060, Fort Leonard Wood, Missouri 65473-8926.

Chapter 11

Closed Circuit Television Systems: Who is There and What are They Doing?

While the choice to install and use a controlled circuit television system (CCTV) is a function of the needs of the facility, the decision to install one should be the result of the risk assessment. In conjunction with other systems, a CCTV system can provide an excellent layer of protection to a facility. CCTV allows surveillance of an area or location and allows managers to observe employees or possible intruders. It can help a facility prepare for a response to threats, incidents of violence or criminal activity. If the CCTV system is designed to record events, a facility can reduce liability because CCTV recordings can provide documented proof of an event. For example, if a customer claims to have slipped and fallen, the CCTV recording of the event may help to prove or disprove the claim.

Installation of CCTV Systems is complex and requires special knowledge and skills. Decisions are required about the location and types of cameras, the wiring between cameras, devices to control the cameras, and the type of recording. In addition, the complexity of cameras requires careful coordination with lighting design. CCTV can be installed inside or outside and the cameras for one location may not perform satisfactorily in the other.

Depending upon the configuration of the total system, a 16 Camera CCTV system with recorders and monitors can be installed for $20,000 to $40,000. A simple one or two camera setup can cost as little as $200. Major systems with dozens of high end cameras, multiple monitoring stations and recorders can cost as much as $200,000.

One of the first elements to examine in deciding to install a CCTV system are the cameras.

CAMERAS

Cameras for CCTV systems come in all shapes and sizes. They can be designed to work outdoors or indoors and they can be overt, where customers and intruders can see them, or they can be covert, where they are unobserved by clients. A facility must decide upon the types of cameras, where to install them and how to wire them. In general a high end camera with pan, tilt and zoom can cost $3000 to $5000 while a button camera can be obtained for as little as $89.

How a TV Camera Works

Before an evaluation of CCTV camera systems can begin, it is beneficial to understand camera basics in order to understand the advantages and disadvantages of the many camera options. For a television camera an image is focused through a lens onto a digital plate inside the camera body. Electrons scan the plate and codes the image into tiny light and dark images called pixels. The signal of the pixels, light or dark, is transmitted through wires (or with radio waves in wireless applications) to another device where the process is reversed and a beam in a television is painted across a screen, illuminating it. This becomes the image that appears on the television screen. How well a system works is a function of the light on the image, the number of pixels scanned by the camera and the ability of the television to project the image. Camera resolution is determined by the number of lines it scans. A camera that scans a lot of lines, 480, has better resolution than a camera with fewer lines, 200. But because of the more lines, more data has to be transmitted.

Motion is captured one image or frame at a time. For cameras the number of frames per second (fps) or images per second (ips) is another element affecting camera performance and price. Most televisions operate on 30 frames per second. A frame speed of 15 images per second is slightly jerky, where some cameras can capture 50, 60, 120 or even 480 frames per second. A camera with a higher frame per second speed that has been recorded can be slowed down when played back. This is how a slow motion camera works. The images are recorded at a fast rate, say 240 frames per second (fps) and then played back at a slower speed, say 30 frames per second. This makes the event seem to occur in slow motion. Slowing the playback down to 15 frames per second gives the images a jerky motion because the changes in movement are too obvious to the viewer. Old Charlie Chaplin movies were shot at a frame rate slower than 30 frames per second which gives the character the jerky, quirky motion that seems so humorous.

As an example of slow motion, many golfers like to have their swing recorded to analyze it. Since the speed of the club during the swing is so fast, a typical camera recording at 30 frames per second will only capture two or three images during the swing. A higher frame rate, say 480 frames per second or images per second will capture more images. Therefore, using a higher frame rate, it is possible to break down the event into more images. For security, using a high speed frame rate can be important because a theft sometimes takes place very quickly. In a casino, the movement of cheating at cards is very slight and very fast. A high frame per second rate is required to catch the exact image.

But the problem with a high frame rate is that it requires the storage of more data. Therefore a high resolution camera (480 lines) with a high frame rate (480) images per second, moves more data than a low resolution camera (80 lines) with a low frame rate (15 ips). The higher resolution and faster frame rate increases the price and size of the camera.

Indoor vs. Outdoor Application

For CCTV systems one of the first decisions to be made is whether to have cameras installed indoors, outdoors or both. For indoor cameras they can usually be mounted in or on the ceiling or they can be mounted on the walls. Fixed cameras can only look in one direction. The fixed dome camera can be placed to view a specific area like a cash register, an aisle in a grocery store, a corridor or a waiting room. Other cameras can pan, tilt and zoom. This means the camera can be controlled to change where it is looking. Pan is the ability to rotate, tilt is the ability to move up and down and zoom is the ability to zoom in closer for a better view. These cameras are called PTZ cameras for short. Figure 11-1 shows a typical dome camera installation in a hospital facility. For indoor cameras the light is under better control than it is for outdoor cameras.

For outdoor use the camera needs to be more robust because of the more extreme weather conditions faced by the outdoor camera. Usually outdoor cameras come in a weather proof enclosure that protects the camera lenses and electronics from dust, snow and rain. Depending upon the design of the housing, it can also protect the camera from heat and cold. As with an indoor camera, outdoor cameras can pan, tilt and zoom or they can be fixed. Some outdoor cameras have a hood that protects the lens from direct sunlight and others have tiny wiper blades that clean the glass. The lens is usually protected inside the closure. Lighting can be a more difficult problem for outdoor cameras since they must operate in daylight and dark. Figure 11-2 shows an outdoor camera.

Figure 11-1. Indoor dome camera. (*Photo by Robert Reid, 2004*).

Figure 11-2. Outdoor dome camera. (*Photo by Robert Reid, 2004*).

Lighting

CCTV Cameras and lighting need to be closely coordinated. The ability of the camera to operate at night is a function of the housing, the lens, the image scanner and lighting used to illuminate the scene.

On a CCTV camera in a chemical processing plant, the camera was enclosed in a weatherproof housing to protect it from wash down solutions when the room had to be cleaned. However, the room was quite warm and the heat from the room plus the internal heat from the camera caused the device to overheat, which made it necessary to turn it off for short periods and allow it to cool. Obviously the overheating of the camera was a factor in reduced camera life. To provide cooling for the camera, instrument air was piped to the housing. The instrument air flow was set so that the air flowing through the housing would cool the camera and prevent downtime. It was simple fix that extended the life and operating period of the $4,000 camera. The elimination of the short outages also improved plant processing capability.

For CCTV cameras that operate inside, lighting is usually steady and is controlled by building lighting. The camera must be compatible with the lighting. The interior lighting is a function of the light bulbs in the light fixtures. Light bulbs can be either fluorescent or incandescent. A facility can have problems if the cameras are selected for incandescent light and the facility later changes to fluorescent lights. Also, to save energy, facilities may elect to change the light fixtures or may change lighting levels at night which will affect the CCTV camera performance. Some CCTV cameras can read both the visible band and the infrared band of light allowing them to see clearly in both light and dark.

Lighting of a scene covered by outdoor cameras changes throughout the day. In a parking lot, for example, lighting in bright daylight at mid-day may change in cloud cover, rain or snow. At dusk less light is available and can be affected by fog. At night lighting conditions change again, and later parking lot lights may come on which would change the scene again. To find a camera that works well in all these different lighting conditions is the goal. Many cameras will work but it may require different lenses or even dual cameras, one to cover during the day in bright lighting conditions and a separate one for night with infrared capability. CCTV system experts recommend facilities insist on a camera demonstration in all light conditions.

For CCTV systems it is usually necessary to illuminate the area being monitored, even if an infrared camera is being used. Infrared illumination can be overt, red lights, or covert, lights with a wavelength invisible to the human eye.

Placement of CCTV cameras outside should also consider placement of the lights in the parking lot relative to the camera location. A CCTV camera that is trained on a parking lot light will have difficulty seeing the ground below, because the light to dark ratio, recorded on the plate inside the camera, is overwhelmed by the bright parking lot light. Another problem with outside lighting is the color of the light emitted by the light fixtures. Some parking lot lights emit colors that tend to wash out the images the camera records. A third problem with parking lot lights is what is called restrike time. Restrike time is the time required for high intensity lights to come on after they have been turned off for a short period, as would happen in a short power outage. High intensity light fixtures, like high and low pressure sodium lights, use far less energy than common incandescent lights and are less costly to operate, but the trade-off in security is that if there is a power outage, the security force will have to wait for a few minutes for the light fixtures to restrike.

Finally, outdoor camera placement must take into consideration the position of the sun at sunrise and sunset throughout the seasons. A CCTV camera that is looking directly into a sunrise or sunset cannot record images very well.

Analog vs. Digital Cameras

Simple CCTV systems use analog signals to transmit the images. An analog signal requires coaxial cables to carry the image to the monitor or recorder. These cameras are less expensive than digital cameras which process the signal and carry it digitally. Digital signals can be carried by unshielded twisted pair (UTP) wire. Currently, most cameras capture the images using analog signals and, for transmission, cameras convert the signal to digital before sending it. The signal can be carried about 400 feet on coaxial cable, and about 1000 feet on unshielded twisted pair wire. Either set of cables can have the signal boosted or amplified to carry it further, but this requires additional equipment.

Color vs. Black and White Cameras

The facility can specify color cameras or black and white cameras. Color cameras generally are more expensive and require more ambient light than black and white cameras. The price and light sensitivity of color systems has been dropping and there are some things that can be seen in color that cannot be seen in black and white. In some instances a facility has been able to upgrade from black and white to color without too much expense and in many cases the cabling can be the same. The price of digital color cameras has been going down steadily in relation to the price of black and white cameras.

Pan and Tilt and Zoom

Regardless of whether the camera is located inside or outside a camera can have the ability to follow or track an individual.

Fixed cameras look only in the direction they are pointed. Some have the ability to see up to 180 degrees because the lens in the camera is a wide angle lens, however, detail is lost with this type of lens. The advantages of the pan, tilt and zoom camera is such that it is controllable from a monitor station. A camera with PTZ can be tilted up to look at the entrance to the parking lot and when a vehicle comes in, pan and tilt can follow the car to the parking place and then zoom can close in on the driver.

Pan, the ability of the camera to rotate, requires a special fitting to allow the camera to rotate full circle. This fitting is expensive and is composed of a sleeve or rings that allow electrical contact connections throughout the full circle of motion. In another application, less costly, the wires to the camera are fed through what is called a pigtail. The wires come out from the camera housing and feed into conduit for transmission. The problem with the pigtail drop is that the camera's rotation with this wiring is limited. Otherwise, the camera could be rotated through several circles and eventually the cable become so fouled that camera can no longer rotate or the motor pulls the wire loose, disconnecting the camera. Since the operator cannot see the cable,

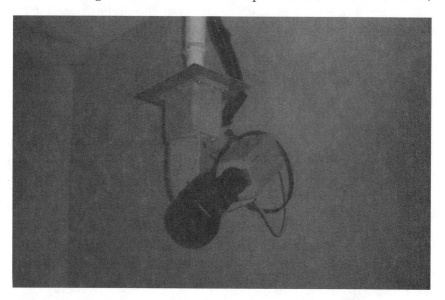

Figure 11-3. Indoor pan, tilt and zoom camera. (*Photo by Robert Reid, 2004***).**

this is a common problem with the pigtail cable mount.

A fouled cable can be awkward to the security staff when it becomes necessary to "unwind" the pigtail to prevent fouling. To have to unwind a pigtail during surveillance is very frustrating.

Tilt, the ability of the camera to look up and down, can also have problems as a result of the way the camera is mounted. The weight of the camera is uneven when the camera tilts up or down, meaning the weight on the mount is unbalanced. Over time this can weaken the mount and can lead to camera instability. Depending upon camera location, tilt above the horizon may not be necessary. The ability of the camera to tilt down to look straight down is usually limited by the mount to something less than straight down meaning the area immediately under the camera is a blind spot. Some installations allow cameras to tilt turn past the vertical which then makes the image on the television monitor appear upside down.

Zoom is a function of the camera's lens. With a better lens, more zoom (close ups) can be performed. One of the problems with zoom is that when a camera is in full close up, jiggling or movement of the camera will make the image bounce. This can happen with outdoor cameras in the wind or when the camera is mounted on a structure that vibrates slightly as when a car moves through a parking garage. The zoom lens also requires more light as the image is magnified.

The control of pan, tilt and zoom is accomplished with tiny electrical motors in the housing/mount. The motors turn small gears that move the lens in/out and move the camera left/right and up/down depending upon the control action. In most cases when the limit of the track is reached the camera will stop because the track has reached a stop. The motor is still trying to turn the camera, but the gears cannot turn further. Continuing to force the camera to try to turn after it has reached its stop can burn out the motor or strip the gears. For these reasons, elaborate cameras have switches that cut out the motor when the camera reaches the end of its track. It won't go further in the direction indicated from the controller. This option increases the cost of the camera and is a possible trade-off in life cycles.

Pan, tilt and zoom are usually controlled from the monitor where the image of the camera is being observed. A control stick is in a box and the joystick is moved up or down, left or right until the view is correct. The control can give two inputs at the same time, up and to the left or down and to the right and the servomotors in the camera accept both signals simultaneously, making a view change easier than tilt left, and then up and then a little more left and a little more up until the object is observed.

Zoom is also controlled from the monitor usually with a pair of buttons, plus for zoom in and minus for zoom out. As with all cameras, pan and tilt should be adjusted before zoom because it can become difficult to find what is necessary to be observed if zoomed in first. A good camera operator can do all three, pan/tilt/zoom, at the same time.

For example, when observing cars in a parking lot, it is necessary to pan and tilt to the car first and then zoom in on it. If zooming in first and the image is greatly enlarged, it can become difficult to find the object without zooming out again to find it.

Finally, if the object is moving and the attempt is being made to zoom in on it and follow its track at the same time, this can be a little tricky. With practice it becomes easier. Sometimes the object moves more slowly than the camera can track. When this happens it is necessary to advance the camera ahead of the object and wait while the object crosses the screen, then move the camera again, and wait again. On the other hand the object can move through the scene faster than the camera can track, meaning the object is lost from view unless the object stops and allows the camera to catch up.

Recently vendors have new software that picks an object to be tracked and the computers will follow the object and zoom in on it, as necessary. This does not work as well as a camera directed by a person from the monitoring station, but it has its applications. This software is not as expensive as one might think; costing less than $4000 and sometimes it is included with a package as a purchase incentive.

The tele-video signal is captured at a certain frame rate (frames per second). If the object is moving faster than can be captured by the frame rate, the object will appear blurry. Also, if the capture rate is slower and the camera is tracked, the scene will be blurry until the camera stops tracking. For these reasons the PTZ cameras usually need a faster frame rate than fixed cameras.

The controls become more complex when attempting to control more than one pan, tilt, and zoom camera at a time. Usually each of the PTZ cameras is addressable. From the monitor, a number is entered, which says, "Camera one, I'm communicating with you." Then the joystick controller directs only that camera. If more than one camera can observe the same spot from different locations, two operators can control two cameras.

However, if more than one operator is trying to control the same camera from two different locations, the control system should be programmed to take commands from the controllers in a hierarchic order. It isn't good for the servo motors to receive a command to pan right from one controller and pan left from the other. The result is a burned out motor and maintenance.

Usually the more expensive the camera's optics and servomotors the longer the camera will last. However, in a CCTV system where each camera has to be maintained and serviced, the facility needs to think about how maintenance on the cameras will be accomplished. If it is critical to capture images, there is not a good time for the camera to be out of order. Most CCTV facilities stock spare cameras so that if a camera fails, it can be quickly replaced without having to wait for a new one to be shipped from the factory.

It is possible for cameras to have zoom without pan or tilt. This allows a fixed camera to zoom in, but it can only zoom where it is aimed. For this reason, only a few of these cameras are installed. In most CCTV security applications, fixed cameras cover the facility with a few PTZ cameras to cover something identified by the fixed cameras in more detail.

Location

The location of cameras is as important as the number and types of cameras. Some facilities have a large number of fixed cameras and few PTZ cameras. Others don't have any fixed cameras and utilize their resources only for PTZ cameras. The decision of where to place cameras is a function of the risks anticipated. If risks are of night intruders, perhaps it is best to locate cameras on the fence line, with a second tier of cameras on the building and finally a few cameras inside to identify the intruders. A camera placed at a higher elevation can see a greater area than a camera at a lower elevation. A camera on or near the floor may be necessary for some special reason, but it won't see much. An overhead camera allows more area to be viewed. Fixed cameras are installed with overlapping fields of view a few feet above the floor level.

For a commercial retail application, it might be more practical to have fixed cameras mounted on the ceiling throughout the store backed up by PTZ cameras in critical locations. Gambling casinos have fixed cameras over every table and the casinos often have additional PTZ cameras to follow individuals through the property. In one instance a jackpot winner was tracked by camera from the payout cage to her car since she had won a large jackpot and the casino wanted to make sure someone didn't try to rob her between the payout cage and the parking lot. She made it home safely with her $2700 winnings.

Cameras inside are usually mounted on the ceiling looking down. Sometimes they are mounted high on walls. Outdoor cameras are mounted on poles or on the roof. Wal-Mart™ mounts cameras on the roof for parking lot security.

Figure 11-4. Outdoor pan, tilt and zoom camera. (*Photo by Robert Reid, 2004*).

Also, if CCTV is located in the ceiling or high up on the walls, as it would be for good visibility, management needs to determine the best method for servicing the cameras for maintenance. The equipment to get to the cameras when they are 30 or 40 feet above the floor may be a costly addition to the maintenance budget.

In a facility constructed of concrete and steel, a block out was put in for a camera and the camera was mounted according to the design. Later the contractor installed a large air conditioning duct near the ceiling where the camera was. The camera could see nothing except the duct. The camera had to be moved to a new location and wire for the camera rerouted. This cost the facility extra money. The HVAC duct installer had no choice where to put the duct.

In most instances CCTV cameras are overt cameras, that is, they are intended to be seen in order to act as a deterrent. In some states it is necessary for the facility to post a notice that the area is being maintained under video surveillance. Some facility managers faced with short budgets install dummy overt cameras that aren't really hooked up. They just look like a camera with a little light emitting diode in the front designed to fool a possible intruder.

Covert cameras can be made to look like anything. Covert cameras can be installed inside a wall clock, a doorbell, a book or a light fixture. These items are usually placed high up on the wall inside a room for more visibility. For outdoor use, covert cameras are usually placed in an enclosure behind smoked or darkened glass. Many ATM machines have a camera to observe the person making the withdrawal behind smoked glass.

The use of covert cameras with signs posted that video surveillance is taking place sends a mixed message. It may be that the facility has covert cameras that work, dummy observable cameras designed to fool intruders, and signs posted that say video surveillance is taking place. Some facilities may have signs stating video surveillance is taking place and have no CCTV system at all.

When a decision is made to install a CCTV system, a facility may post a sign indicating there is video surveillance before the camera system is operational. This acts as an added deterrent.

Camera Maintenance

As with all security systems, maintenance on the CCTV cameras is a necessity. Maintenance includes checking each device to confirm it is working correctly, cleaning lenses and housings and lubricating the servo mechanisms on the pan and tilt cameras. It helps to routinely check and tighten the electrical contacts of the cameras as well. The average life of a CCTV camera is between 15,000 and 40,000 hours. At that point, it may be beneficial to replace all the cameras at once, rather than letting them fail one by one and having technicians change each one out incrementally. The state of the art of cameras is advancing rapidly and cameras even 5 or 7 years old can be replaced easier than repaired. Camera internals are like all electronic parts. It is usually easier to replace them than to repair them.

While the camera is one of the most important parts of the CCTV system there are other elements of the system that must work together if the system is to perform as required.

MONITORS

In addition to cameras, a CCTV system requires television monitors for staff to view the images captured by the cameras. For a simple system one camera can send a signal to one monitor. However for systems with many cameras it isn't practical to have an equal number of monitors. So most CCTV systems have fewer monitors than cameras and use a device called a multiplexer to allow operators to select which signals from the cameras they want to observe on the monitors.

Monitors can receive signals from several cameras at the same time with a feature called split screen. Up to sixteen cameras can be viewed with a single monitor, but in this case each image is quite small. Many installations have a four camera system and use it to feed one monitor that has a 17 to 20 inch screen. A 16 camera system usually has 4 monitors. A monitor is very similar to the television set except the signals that are fed to the monitor come from the dedicated CCTV circuits instead of from the broadcast circuit.

Black and White vs. Color Monitors

Even when the CCTV system has color cameras, it is not necessary to provide color monitors. The monitors can be either color or black and white. In some cases, a facility records color images and uses black and white monitors for viewing. Later the facility can use a color monitor to view the recordings if something important has been recorded. Generally, color monitors are larger than black and white monitors and this can affect the layout of the control console. In addition, color monitors are more expensive than black and white monitors. Each type of monitor has about the same service life.

Monitor Size

Most facilities have monitors that are desktop sized. Usually 12 inches to 20 inches, measured diagonally. Some control stations will have 4 monitors with each monitor displaying the output from 4 cameras. So a 16-camera system would have 4 monitors set in a bank of 4.

Figure 11-5 shows a control station with 12 monitors, 9 are black and white and three display color. Some large CCTV systems can have a large monitor on the wall of the control room; say 50 or 60 inches diagonal, while smaller monitors can be seen from operator consoles. Some CCTV systems have very small monitors set in a bank or rack. This means each monitor is displaying the signal from one camera. These small monitors can be 5 or 6 inches diagonal in size. While this type of system was popular in the past, it isn't seen as often today because, with the larger monitors displaying multiple cameras, the display can be switched to show the output of one camera full screen. This gives the operator the advantage of more detail for close surveillance, while the monitor is set to multiple camera display for routine surveillance.

Monitor Controls

A CCTV monitor is very similar to a television. It has channels, one for each camera, and it has the standard vertical and

Figure 11-5. CCTV camera monitoring station. (*Photo by Robert Reid, 2004*).

horizontal hold. Most CCTV systems do not include an audio signal so there is no sound or volume controls. The monitor includes an on and off switch and it may be a matter of policy to turn off the monitors when visitors enter the area. Other controls include sharp and bright for black and white monitors and tools for adjusting the red, green and blue images for color monitors.

In addition to the controls to view the image, there is usually a device for controlling the pan, tilt and zoom for the cameras. This includes knobs or a stick controller to control the pan and tilt and buttons for controlling zoom in and out.

The channel changer is usually different from a television. Depending upon system design, a device called a switcher or switch is installed between the cameras and the monitor. Sometimes the switch comes with the monitor.

In an analog system, the coaxial cables from each camera are fed to the switch and a single cable is fed from the switch to the monitor. Calling up a camera is done by entering the number of that camera. Most control stations will have a map of the facility with cameras marked on it as a display. The number on the map indicates the number of the camera. Then if the operator wants to view the main entry, he looks on the map, sees that it is camera 3, for example, and enters "3" into the monitor. The image of the main entry appears on the monitor.

Computer Monitors

Some facilities are able to send their camera images over an internet protocol and provide a signal through a local intranet or the internet. Anyone with a computer and access to the network can call up images. The same control functions in the control room can be performed by the computers, camera choice, split screen, pan, tilt or zoom. However, the security organization of many facilities keeps the CCTV monitor separate from other computers by giving the staff both monitors and a computer. Otherwise, someone reading emails, or updating access control databases for entry can miss viewing images on the CCTV monitors.

Staff Use

While Chapter 15 addresses itself specifically to security and personnel matters, it is important for facility managers to recognize that staring at a CCTV monitor, or bank of monitors for long periods can be tedious and eye straining. For this reason, monitors are usually kept dim, and the room where the monitors are viewed is also dimmed to make observation of the monitors easier. Frequent breaks from monitoring viewing are also necessary.

The CCTV system starts with a single camera and monitor, but it quickly becomes more complex with multiple cameras and monitors. Monitors are an important part of the system because staff can view what is happening in front of the cameras and take action against intruders, if necessary.

RECORDERS

With the cameras in place and monitors to see what is going on, the next tool in the security manager's CCTV system is the recorder. It is possible to record the images viewed by the camera. With a recording, it no longer requires the eye witness testimony of the person who viewed the event. A recording of the event can be presented. This is an extremely valuable tool to the facility. CCTV images can be recorded digitally or on VHS recording tape.

However, one can easily see that recording the images of 16 cameras or more for 24 hours per day eventually amounts to a huge amount of data that needs to be stored. A facility manager can't afford to have 16 video recorders running all the time, changing tapes every two hours day after day. And what would one do with all the video tapes? If there are no intruders there are hours and hours of a video of a door or window or fence with nothing.

Digital Video Recorders

The digital video recorder is becoming the device of choice for security system applications where recording of the images is desired. A digital video recorder is, in effect, a computer with a large hard drive that stores images in video format. Depending upon the size of the drives and the multiplexers, digital video recorders (DVRs) record the images from multiple cameras. Video format for computer images varies and the facility must decide the format to use to store the images. Video images are stored as files of short duration on what is called "capture cards."

The capture card is an important part of the digital video recording device. Like computers the capture card is usually manufactured separately from the recorder. A recorder manufacturer purchases the card from a card designer. The capture card performs the important function of taking the analog signal from a camera and converting it to digital signals, then compressing the information so that it can be stored on the digital recorder's hard drive. Different capture card designers utilize different protocols for storing the images.

Video Recording Protocols

Several competing compression technologies exist for storage of video motion and some of the more sophisticated computer users (or web developers) may be familiar with the following terms. Video motion can be stored as JPEG, Wavelet, MPEG, and H.263 protocols. JPEG and Wavelet protocols are known as whole image formats, where the whole image is stored every time. (Remember the discussion of 30 frames per second up to 480 frames per second?) Other protocols like MPEG and H.263 do not store the whole image every time. Instead they only store what has changed from the previous image. MPEG and H.263 protocols reduce the size of the video files which allows greater storage on smaller drives. To combat this technology DVR capture card manufacturers have developed "motion JPEG" or "modified Wavelet" as a way to enhance compression of files under these protocols.

Another way of compressing the data is to utilize digital motion detection to trigger recording. The camera captures the images and they are converted to digital signals. Then each frame or image is compared to the previous one. If nothing has changed, the images are not recorded. But if the pixel images change from one frame to the next, a signal is transmitted that tells the recorder to start recording the images. This technology allows selection of a part of the image or the entire image. In an outdoor application for example, the motion detection can be trained on the door. If the door opens, recording starts. However, if a person walks past the door, even though their motion is "seen" by the camera, because they don't move into the part of the image by the door, their motion is not recorded. Some video motion recorders provide buffering capacity. That is, for thirty seconds the images are recorded, and when motion is detected this previous 30 seconds is also written off to the recorder. Generally, video motion detectors are limited to indoor applications because clouds passing over the area or animals moving, winds moving grass or trees, or water, will cause recording to start.

In addition to concerns about compression protocols, it is important to the facility to understand the size of the images converted. Images that are 320 X 240 pixels are smaller than images that are 640 X 480 pixels. The images can be blown up to a larger size, but resolution is lost as the size is increases because each pixel is enlarged. If the images the facility wants to capture are small, like playing cards on a blackjack table, the smaller images may not be able to detect the small subtle hand movements. The whole picture can be seen, like how many players and how many cards, but in a card game, only a very small part of the card is shown when dealing off the bottom of the deck for example. So the size of the image (in pixels) is important and must be weighed against the need to capture the entire scene.

CCTV has had difficulty standardizing on a common resolutions chart and there are different resolutions for color vs. black and white images.

The capture cards in some recorders and cameras allow the user to choose a favorite protocol. As the industry continues to grow, motion picture protocols may settle somewhat. But for now there are many types. Given that this type of issue is developing along with the rest of the computer industry, a facility can be assured that the technology for motion picture capture protocols is sure to change. At some point in the not too distant future, the technology chosen by the facility may become out of date.

Unfortunately there is no sure way to determine which technology will emerge. But many systems using out of date technologies are still being used in other industries. So, just because a technology is out of date, does not mean it will not work, it just will not be as flexible or work for as long as a system based upon a newer technology. For this reason some facilities are continuing to use video tape recorders based upon VHS format.

Once the capture and compression technologies are determined the next important element in DVRs are the size of the storage drives. These have been 160 gigabyte drives but the state of the art is moving to 250 gigabytes. And the storage capacity of these devices increases every day. Along with the size of the hard drive, the choice of a DVR can be affected by the amount of random access memory (RAM) present. With more RAM more images are stored into memory before the machine has to write it off to the hard drive. Since the drives are large, and so much data is being stored, the heads on the hard drive are subject to a higher "duty factor" if the unit has a smaller amount of RAM. This means that the head that reads and writes from the hard drive seeks and stores information more often. Hence the hard drives in digital video recorders for security service need to be more robust than standard commercial hard drives.

Finally the DVR operates with computer software that tells the system what to do. The two most popular software packages are Windows™ and Unix™. Both applications have worked well; however, Windows™ has had problems with some of its applications which were to be fixed with the issue of Windows XP®. Current information on digital video recorders has not defined whether these concerns have been resolved. Unix® has a public domain system called Linux® which has an attractive price but many devices don't provide drivers that use Linux as a base. This causes Linux to have limited application.

Some DVRs come with on-board multiplexers and monitors which eliminates the need for separate monitors. Some have demountable hard drives which allow switching of hard drives while viewing the captured video on another system. Other DVRs have controls to send out signals to PTZ cameras for control.

Digital video recorders cost between $700 and $5000 dollars. The less expensive ones have less storage and record fewer channels (cameras). Less storage means the length of time to be stored is shorter. Most facilities elect to store images for 30 days.

VHS Tape Recorders

Less expensive than the digital video recorder is the VHS tape recorder. These devices are still successfully used for many facilities and they are less expensive than their digital video recorder counterpart. A VHS recorder designed for a security application that is capable of 960 hours of storage in around the clock operation (24 hours per day) costs about $375. This machine is capable of recording about 40 days worth of information. VHS recorders are capable of recording up to 16 cameras at a time. In addition to the recorders, tape systems require tapes to store the information. Blank VHS tapes are easy to purchase. Special application tapes for VHS applications are not required. Blank VHS tapes can be purchased for under $5 apiece.

VHS tape machines are fairly reliable, but eventually the heads wear down and require cleaning. The life of a VHS recorder being used for taping video surveillance is usually less than 7 years.

MULTIPLEXERS/SWITCHERS

The final element common to most CCTV systems is what is called the multiplexer. A multiplexer takes the signals from several cameras and combines them into a single data stream. This way the signal of multiple cameras can be sent over one base cable. At the other end, another multiplexer unscrambles the television signals allowing any one of the cameras to be viewed on a monitor. Multiplexers usually receive the input from coax cables from each camera and can then transmit the combined signal over either coax or unshielded twisted pair wire. Some multiplexers are capable of transmitting the signals on fiber optic cable. Multiplexers are sized for in and out, for example 16 and 2 which means there are 16 inputs and 2 outputs. This would mean that two monitors in the control room could look view any of the signals from any of the 16 cameras.

A 16 and 1 multiplexer could capture the signals from 16 cameras, and transmit it to another 16 and 1 multiplexer where the signals are broken down at the other end. In this way, only one cable is required from the multiplexer to multiplexer rather than run sixteen coaxial cables, one from each camera to a monitor.

The weakness of this type of system is in the single cable between multiplexers and in the multiplexers themselves. If the cable or one of the multiplexers fails, all camera data is lost until the multiplexer can be repaired.

INTERNET PROTOCOLS

As more and more cameras become digital cameras instead of analog cameras, and as broadband capacity increases, it has become possible to have one circuit or loop running through the facility with all the devices clipped onto the circuit. Each camera and monitor has its own address. In this instance, a camera with an Internet protocol address is hooked to the network and in another place a computer or monitor with another address is also hooked to the same loop. This allows any computer or monitor on the loop to look at the output of any camera. In some ways this is what happens with the weather cam cameras that are currently on the internet. If you know the correct web address you can call up that camera and look at the live camera feed to see what the weather is. For security purposes this might be a little too wide open but the technology is certainly there.

For some systems, it is possible to have access available only on the local area network. For security proposes, a double loop might need to be provided. In this way, the second loop backs up the first by providing redundancy if one of the loops fails.

CASE STUDY

A facility used CCTV cameras to observe plant operations at night. Using multiple cameras it was possible to observe plant operators in various areas without making the observation known. It was also possible to check several areas using multiple cameras and to record the output of one or more cameras on video tape. Special work was always recorded as a precaution. All cameras at the facility were PTZ cameras but the CCTV system was primarily an aid to operations and not used for security. Use of the cameras allowed work to be observed from the control station rather than at the work station. Each camera cost $6000 dollars and the facility had approximately 45 cameras. The total cost of the system including the cameras, multiplexers, monitors, recorders and cabling was $420,000.

RESOURCES

Ohio, Denise *Five Essential Steps in Digital Video: A DV Moviemaker's Tricks of the Trade* Que Corporation, 2002.

Chapter 12

Eavesdropping & Voyeurism
And What to Do About It[1]

SUMMARY

News reports about spy cams, bugs and wiretaps found in commercial facilities are increasing. Increases in electronic surveillance detection inspections within facilities parallel this trend. In fact, there is a good chance you are reading this because someone came to you and said, "I want the place swept for bugs." You need an instant education in order to respond quickly and competently. This chapter fills that void. When initially approached by management with their "strange" request to debug or sweep the facility, you might think they are just being paranoid. They are not. The thought would not have occurred to them if everything was fine. Something is wrong. Something valuable is at risk. Take the request seriously. Enlist the aid of the best counterespionage consultant that you can find. Later in this chapter I will show you exactly how to do this. Hint... forget the Yellow Pages, your security guard company, and general private investigators.

Before you do anything, knowing why you are doing it will add purpose to your mission. Ask a few questions...

What is an inspection for electronic eavesdropping?
Inspections are a defense against...
Competitive Intelligence
Corporate Espionage
Terrorism

---

[1]This chapter was written by Kevin Murray, CPP, CFE, BCFE, of Murray and Associates at www.counterespionage.org.

Malignant Activism
Extortionography
Internal Intrigue
Strategy Spying
Media Snooping
Mysterious Leaks
Personal Privacy Concerns

What do inspections protect?
Regular eavesdropping detection inspections protect...
Sensitive Communications
Boardroom Discussions
Mergers & Acquisitions
Delicate Negotiations
Lawsuit Strategies
Employee Safety
Trade Secrets
Personal Privacy
Vulnerable Off-site Meetings
Executive Residences & Home Offices

What do routine inspections accomplish?
Accomplishments include...
Discovery of covert surveillance devices (of course)
Identification of other security loopholes
Helping to fulfill top executive's fiduciary responsibilities
Helping to fulfill due diligence requirements
Helping to fulfill "Business Secret" requirements for court

CREATING THE EAVESDROPPING DETECTION INSPECTION PROGRAM

Although your initial inspection might be one which was initiated by a specific problem, lasting protection comes from instituting an on-going program. Most companies now think of it as just another element in a complete security program.

Assume you have just been asked to look into this facet of security and were requested to report back. Your boss asks, "What is a surveillance inspection program?"

You answer, "Electronic eavesdropping inspections—sometimes referred to as technical surveillance countermeasures, or TSCM, are a company's systematic effort to detect intelligence collection efforts. They provide us with early warning. Thus, we can thwart their efforts before we are harmed. Very cheap. Very effective."

"What areas should be inspected?"

"We don't need to inspect everything, just our sensitive areas: offices, conference rooms, executive dining areas, off-site meeting locations, executive homes, vehicles, etc."

"How often should we inspect?"

"The generally accepted corporate and government security practice is quarterly inspections; with bi-annual inspections being the next most common schedule. Of course, rooms with different sensitivity levels may be mixed or matched for maximum effectiveness. In times of crisis or heightened sensitivity, more frequent inspections will be required. Occasionally, there may be an area where an annual inspection will suffice. I will work all this out with our counterespionage consultant."

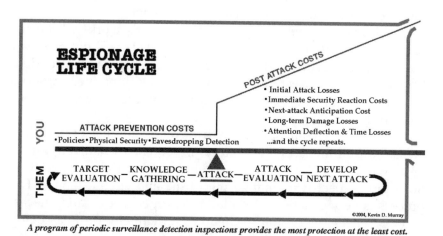

A program of periodic surveillance detection inspections provides the most protection at the least cost.

Figure 12-1. Espionage life cycle. Source: Murray and Associates, Oldwick, NJ.

Why do inspections need to be repeated?
To detect the signs of intelligence gathering or employee snooping—a prelude to the real trouble.
To allow time to counter before any real harm can be done.
To limit windows-of-vulnerability.
To satisfy on-going legal requirements for due diligence.
To help satisfy the requirements for attaining "business secret" status in court.
To identify new methods of privacy invasion and information theft.
"And you know, not reinspecting periodically may be interpreted (in court) as an admission that previously protected information is no longer important."

The next logical question…
"How are you going to do this?"
Start your program, or upgrade your current efforts, by working with an independent security consultant who specializes in surveillance detection testing and overall counterespionage consulting. Make sure they are not product affiliated. You need to be able to trust their judgment 100%. Accepting remuneration from product sales, accepting commissions or kickbacks, or having a menu of "other recommended services" that their company can provide, clouds judgment.
The most visible part of your specialist's work will be the search for electronic eavesdropping devices. This is not however the only part. A good counterespionage consultant will also identify, and make corrective recommendations, for other info-loss vulnerabilities as well, e.g. inadequate perimeter security or poor security habits.
Work with your specialist to develop this list of sensitive areas. Not all areas of the facility are equally sensitive. Not all areas will always be sensitive. The list of areas to be inspected is not rigid. Expect it to change with time and circumstances. The process is very economical if planned properly. It is rarely necessary to "check everything." Creating a hierarchy of areas being inspected increases effectiveness and reduces costs.
Once the sensitive areas in your facility are identified, your specialist will help you decide how often to conduct

reinspections. Each area will have its own window-of-vulnerability-tolerance (WOVT).

How do I identify the right specialist for my needs?

Find an independent security consultant who specializes in electronic eavesdropping detection and espionage prevention. Recognizing this person may not always be easy. There are too few of them, and too many pretenders.

Contact your fellow facilities managers and corporate security directors. Ask them who they are using. Get first hand recommendations. You may also seek out consultants via industry association membership directories:

International Association of Professional Security Consultants 515-282- 8192 http://www.iapsc.org

The Espionage Research Institute 240-273-8823 http://www.espionbusiness.com

Conduct an Internet search using key words like "eavesdropping" or "counterespionage." Carefully evaluate (very carefully) the web sites you come across. Red flags include little, if any, information about the people who will have access to your sensitive areas; and self-aggrandizing statements that sound too good to be true. Always ask for references. Always verify. Always.

Before developing a long-term relationship with any specialist ask lots of questions...

Is electronic eavesdropping detection your company's only business?

How many years have you been specializing in eavesdropping detection?

How many years under your current company name?

What did you do before that?

How many electronic eavesdropping investigations have you personally handled?

Are your recommendations independent and unbiased? Do they profit you—in any way?

Is your firm licensed by a state police authority?

Are resumes available for every technical investigator who will be on my premises?

Can you provide copies of your certificates of insurance for me?

Do you have any professional certifications? (CPP, CFE, BCFE...)

Do you teach security or counterespionage on a university level?

Have you written books, chapters or articles that have been published? May I see them?

Do you have references? (Be sure to call them.)

What is the dollar value of the instrumentation you use?

Do you own all your own instrumentation? (Do you: rent; or "borrow" from your full-time employer?)

What is the average age of your instrumentation?

Will you allow your eavesdropping device 'finds' to be verified via polygraph testing?

Will you sign a "confidentiality agreement?"

Are you qualified to investigate general corporate espionage too?

In how many court cases have you been qualified as an expert witness? (List please.)

Do you use some form of radio-reconnaissance spectrum analysis (RRSA)?

Do you use non-linear junction detection to find dormant bugs?

Do you use time domain reflectometers to check every inch of phone wiring?

What is your wireless LAN inspection procedure?

Is Thermal Imaging Spectrum Analysis® (TESA) part of your procedure? What camera do you use?

How will you inspect for infrared emissions, fiber optic devices, etc.?

How will you customize your procedures to meet my needs?

Do you charge extra for inspecting on nights, weekends or holidays?

After you have the responses to these questions, ask yourself...

Are they sincere about solving my concerns, or are they money driven?

Will their presentation, dress and demeanor reflect well on me in my boss's eyes, or in court?

Trusting my instincts, do I really feel they are capable of solving the concerns here?

If you are still unsure after all this, keep looking. Success hinges on finding the right firm or specialist.

Resist the temptation to bypass the right firm or specialist in lieu of someone who is almost as good just because there are some travel expenses involved, or their fees are lower. This is false economy. Engage the very best person you can find. You may only get one chance to "do it right." Remember, fees and expenses are minor compared to the value of what you are protecting. Do your research. Pre-qualifying your eavesdropping detection specialist is important.

> Recall the discussion of performing the risk assessment from Chapter 1? If you have your risk assessment, then you can compare the value the assets to the cost of the eavesdropping surveillance. Don't forget to include liability for loss of critical information.

WHAT IS THE AVERAGE COST OF AN INSPECTION?

Answer—"A whole lot less than suffering a loss or a lawsuit."

Eavesdropping detection inspection services are usually charged on a per-item basis, or in the case of smaller assignments, a pre-agreed upon flat rate. All budgets may be accommodated simply by inspecting the most sensitive areas first. Counterespionage consulting is usually billed on a daily basis.

Per-hour fees are no longer used by true countermeasures specialists. The length of time it takes to complete an inspection has little bearing on how effective the search is. You are paying for value, not time. A knowledgeable and well-equipped specialist, for example, may complete an inspection in half the time required by someone who is ill-equipped and not very knowledgeable. A qualified specialist will also be many times more effective.

Shopping Tips...

Always ask to see a written Fee & Expense Schedule before discussing job details.

Avoid per-hour pricing.

If you hear "I'll have to stop by your location first..." instead of receiving the straightforward pricing you requested—you're probably in for a high pressure sales call.

Inadequate pricing cheats you of quality expertise and instrumentation. If the prices seem low, suspect that the person has not invested in quality instrumentation, training and insurance. If the cost is unrealistically high, you can rightfully suspect that this person doesn't have much field experience either.

The following samples reflect the charges required to provide corporate/government level eavesdropping detection services. The average prices shown are current as of Fall/Winter 2004/2005.

Small inspections...
Five average size offices.
Five telephones, associated wiring and switching equipment.
Final written report.
Approximate cost: $3,700.

Medium inspections...
Eight average size offices.
Eight secretarial areas.
One large board room.
Sixteen telephones, associated wiring & switching equipment.
Two fax machines and associated wiring.
One speakerphone system.
One video teleconferencing unit.
Final written report.
Approximate cost: $8,500.

Large inspections...
Twenty average size rooms.
One large board room.
Forty telephones, associated wiring and switching equipment.
Five fax machines and associated wiring.
Six speakerphone systems.

One video teleconferencing unit.
Includes: Final written report.
Approximate cost: $18,500.

INSPECTION PROGRAM

"So tell me, what is your inspection program?"
You should hear something like this...
"We come to your facility with our instrumentation at a mutually convenient time. This visit is usually scheduled during off-hours so as not to be disruptive to your work. The actual inspection happens something like this..."

Preliminary Evaluation

At the outset, the specialist conducts a background interview to obtain an overview of the security climate, concerns and culture. (This discussion is **not** held within the areas being inspected.) Just like a doctor, they want to fully understand the symptoms and circumstances that preceded the call for assistance, or the decision to begin a pro-active protection program.

Survey of Current Security Measures

This includes an inspection of perimeter and interior physical security hardware. Doors, locks, windows, vents, alarm devices, wastepaper disposal methods, etc. It also includes a review of current security policies and procedures. A tour of the facility may be part of the process. Have all the necessary keys available, and if necessary, a copy of the floor plans.

Visual Examination

The areas in question are visually inspected for eavesdropping devices, and evidence of prior eavesdropping attempts (bits of wire, tape, holes, fresh paint or putty, disturbed dust, etc.). The technical investigators rely heavily on their eyes, knowledge and experience during this stage of their work–these are the finest detection instruments available. The visual inspection is thorough and includes: furniture; fixtures; wiring; ductwork; and small items within the area.

Acoustic Ducting Evaluation

Unexpected sound leakage into adjacent areas has been found to be the cause of many information leaks, especially the in-house type. Open air ceiling plenums, air ducts, common baseboard heater ducts, walls common with storage/rest/coffee rooms, and holes in concrete floors have all aided eavesdroppers at one time or another. The acoustical ducting evaluation takes all of this into consideration.

As you see, electronic eavesdropping detection begins long before the specialized electronic instruments are unpacked.

Inspection of Telephone Instruments

An extensive physical examination of telephone instruments is undertaken. There are many types of attacks involving bugs, taps, and wiring modifications which can compromise a basic telephone instrument. Business telephones have additional vulnerabilities, some of which are legitimate system features that, when abused, become eavesdropper-friendly.

After a telephone instrument is opened for inspection, it is put back together and its screws are sealed over with security tape. This provides visual proof that the phone has not been opened since the technical investigator last inspected it.

Security seals should be custom-made and serialized so that they cannot be duplicated. Computer printed sticky labels, nail polish, or even stock security seals are not adequate in this situation.

You can periodically inspect your consultant's security seals yourself. Broken seals may indicate an intrusion, while missing seals may indicate a switch of telephone sets. Either condition is suspicious and should prompt a call to your specialist (from a safe phone, of course).

Inspection of Other Communications Devices

Other communications devices like: faxes, speakerphones, modems, computers, etc. are included because they may carry information the eavesdropper finds interesting. One not-so-obvious reason for inspecting these devices is that their connections to the outside world can be hijacked. Standard audio and video room eavesdropping devices just love fax and modem lines,

LANs, VoIP (Voice over Internet Protocol) and wireless LANs. All are additional sound/video/data-moving conduits which need to be inspected.

Inspection of Telephone Wiring

Wiring associated with the telephones under test is inspected for attachments and damage. Damaged wiring is often the only evidence of a prior wiretap. Junction blocks-where telephone wires connect to each other within a building-may also be inspected. These connected wires form a path between the telephone instrument and the on premises, telephone switching equipment. In some cases (e.g.: simple residential phone service and facsimile machines) internal wiring connects directly to outside cables which lead to the phone company central office. Junction blocks are an easy and relatively safe place to attach a wiretap device. Extra wiring paths may also be constructed at junction blocks (using the spare wiring already in place) to route the audio/video/data to a remote relay device or a listening post.

Telephone Room Inspection

The building telephone room houses more junction blocks for the internal phone system; switching equipment for the internal telephone system; and telephone company junction blocks for the incoming lines. This is another area of vulnerability which requires an inspection from both a wiretapping and physical security point of view. In large buildings, this room is usually found in the basement/utility area. Historically, small to medium-sized telephone rooms have received minimal security attention.

Phone Line Electrical Measurements

Measurements are taken and compared against telephone industry standards. Readings which deviate from the norm can help reveal certain types of wiretaps.

Time Domain Reflectometry Analysis

In this test, a pulse is injected into the telephone line. If the two wires are intact and parallel to one another, the pulse continues its trip smoothly. If the pulse passes a point where these is a change in the wiring (splices to other wires, a wiretap, a wall plug, the end of the wires, etc.) a portion of the pulse is reflected back and alerts the technical investigator to a possible problem.

An instrument called a time domain reflectometer (also known as TDR or cable radar) injects these pulses, reads their reflections, and measures the time difference between the two events. This allows the TDR to calculate the distance to the irregularity. A time-verses-irregularity graph is displayed on the TDR's display. This signature is interpreted. Imperfections in line integrity are calculated to within a few inches of their actual location. An in-person inspection of these points is then made. This allows a thorough examination of the wiring even when hidden from normal view. Time domain reflectometry allows reliable testing of phone wiring up to 2,000 feet, and detection of some wiretap attacks at distances of up to 36,000 feet.

Non-linear Junction Detection (NLJD)

This detection technique, similar to retail shoplifting tag detection, is used to locate the semiconductor components used in electronic circuits, e.g. diodes, transistors, etc. Bugging devices

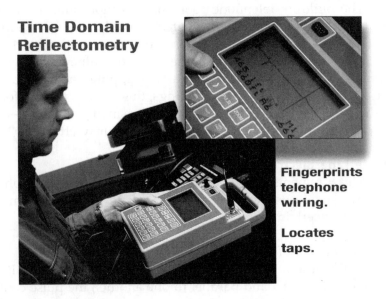

Time Domain Reflectometry

Fingerprints telephone wiring.

Locates taps.

Figure 12-2. Time domain reflectometry. Source: Murray and Associates, Oldwick, NJ.

which contain these components (transmitters, tape recorders, amplified microphones, miniature TV cameras, etc.) are discovered in this manner. They are detectable even when secreted inside walls and objects by using an NLJD. The NLJD emits a radio signal and listens for the return signal from the electronic parts which make up eavesdropping devices.

A special feature of the NLJD is that location of the device is not dependent on the device being active at the time of the search.

RF Spectrum Analysis/Radio Reconnaissance Spectrum Analysis® (RRSA)

Eavesdropping devices which transmit a radio signal (over-the-air or on building wiring) may be detected with an instrument called a spectrum analyzer. In simple terms, this device can be thought of as a radio which has a very long and continuous tuning dial. The received signals are shown on a display screen for visual analysis, and are also converted to sound. Radio reconnaissance spectrum analysis is a technique which carries the detection process several steps further.

Each signal the technical investigator receives is evaluated to

Figure 12-3. Radio frequency spectrum analysis device. Source: Murray and Associates, Oldwick, NJ.

determine if it is carrying voice, data or video information from the sensitive areas being inspected. Analysis also includes converting video signals into viewable and documentable television pictures. Capturing eavesdropping evidence on the-fly is quite important, but may not be available from inexperienced or under-equipped purveyors.

In addition to detecting video bugging devices, the RRSA technique detects computer emissions. These are signals inadvertently emitted by some computers which can be received and reconstructed from a considerable distance away. The technique also detects emissions from computers which have been deliberately bugged.

Due to the sensitivity of an RRSA system, radio transmissions from bugging devices are detectable even if the device is not in the vicinity of the areas being inspected. This means that although only certain rooms may be slated for inspection, entire sections of buildings benefit from this particular test.

Thermal Emissions Spectrum Analysis® (TESA)

Electronic eavesdropping devices and covert spy cameras are discovered with speed and certainty thanks to a relatively new detection method: Thermal Emissions Spectrum Analysis® (TESA).

TESA allows hidden bugs and spy cameras to be "seen" on a portable video display by virtue of the minute amounts of heat radiated as electricity flows in their circuitry. Surveillance devices hidden in ceiling tiles, in walls and in other common objects create slight warm spots.

Detecting eavesdropping devices requires sensitivities in the thousandths-of-a-degree Celsius range—much less than the amount of heat your fingertip leaves on an object after you touch it for a split second.

Currently, thousandths-of-a-degree level of sensitivity is only available in special lab-quality instruments priced in the $50,000 and up range. Lab-quality TESA instrumentation is different from the utility-grade infrared cameras used by police and electrical inspectors—as different as prop planes are from commercial jets. Utility infrared cameras cost $8,000-$30,000, but their sensitivity is only in the tenths of a degree range.

The camera and battery hidden behind a ceiling tile glows when viewed using Thermal Emissions Spectrum Analysis.

Figure 12-4. Thermal Emissions Spectrum Analysis® Device. Source: Murray and Associates, Oldwick, NJ.

Availability of this very worthwhile test procedure is still limited due to the cost of instrumentation. Try to select a consultant who employs this level of testing. Although it is not the only indicator of professional competence, the ability to deploy the latest detection technologies is a good start.

ADDITIONAL TESTS

As in the medical profession—counterespionage consultants also have many tests that are selectively applied depending upon a client's specific needs or concerns. Every situation is a bit different.

In addition to the group of inspection procedures already mentioned, there are tests which are used as the situation demands. A good *technical investigator* will bring quite a bit of additional analysis and thought to the inspection process. The overall goal of your specialist should always be *solve the concern*, not simply dash blindly through a one-size-fits-all checklist. "Checklist Charlies" are a spy's best friend.

Expect to be taken on a guided tour of the whole inspection process, test by test, (in easy to understand terms) the very first time one is conducted for you. The more you know about what the *technical investigators* are doing, the better it is for all concerned.

In addition to the tests outlined above, the investigative process should also take into consideration infrared, fiber optic, hydrophonic and new eavesdropping threats—which will develop in the months and years to come. New threats always arise with new technology.

FINAL REPORT

When your inspection is complete, you should receive a full verbal debriefing. In this meeting the lead technical investigator highlights all serious problems found, and recommends solutions which may need to be implemented promptly. Expect a detailed written report within a week. A final report should include...

A statement about why the inspection was undertaken: proactive or active problem.

A description of all the areas and communications equipment inspected.

An explanation of all tests conducted.

The findings.

Recommendations for security improvements.

A review of other espionage loopholes found.

Security improvements since the last inspection.

Photos, floor maps, inspection history logs, etc.

And other useful espionage prevention information.

Final reports are important documents. Safeguard each one. Together they show a continuing effort to provide information security for specific areas within your business or agency. This is proof that you took extraordinary steps to legally classify your information as proprietary and secret. You have gone above and beyond LAG (locks, alarms and guards). Courts will now listen, stockholders should be satisfied, and snoops will have to move on to someone else's door for easier pickings.

IMPORTANT EXTRAS

A counterespionage consultant would be seriously remiss if *only* electronic eavesdropping issues were addressed.

Experience has shown that few information leaks can be blamed solely on electronic eavesdropping. Sure, eavesdropping may be the most devastating form of espionage—that information is the freshest. But this is only one piece of the puzzle. To see the entire picture, a good spy will collect the other puzzle-parts as well. Each part may seem innocuous in and of itself, but they are synergistically related.

Make sure your specialist takes a holistic approach to information security and endeavors to solve your problems or concerns no matter what the actual cause.

Trust Your Management's Instincts

When it comes to eavesdropping, the thought would not have occurred to them if a problem did not exist. Tell your specialist whenever an executive suspects an information loss. This is a legitimate warning flag. It is not paranoia.

Only Failed Espionage Gets Discovered

You never hear about successful eavesdropping or espionage attacks. You're not supposed to. It's a covert act. Frequency of publicity is on par with commercial airline flights: only the partially completed flights (crashes) make the news. Watergate, for example, was a classic case of espionage incompetence in action. But, for every Watergate, there are many silent successes.

This apparent quiet is what gives uninformed people the feeling that spying doesn't occur—it is a false sense of security. Not only is information theft invisible and silent, it is also prevalent. Spying is a common activity. Discovery relies heavily on proactive inspections-and your intuition.

Due to the covert nature of spying, the exact extent of it is not known. However, we can use the failed espionage attempts as a gauge. They reveal over and over again that the problem does exist. Also, the plethora of electronic surveillance equipment being openly sold in "spy shops" over the Internet and in "executive toy" catalogs gives us a good indication of the magnitude of electronic eavesdropping.

Documented cases of eavesdropping and espionage appear in the news regularly. They show that, left unchecked, spies and enemies can desiccate a bottom line, wipe out a competitive advantage, and leave a company a shell of its former self.

Even more embarrassing is the proliferation of cheap mini-video cameras planted by voyeurs. If your customers or employees are victimized in their birthday suits—and you haven't conducted counter surveillance inspections—you will find a lawsuit not quite as surprising.

Espionage is Preventable

Information is like any other corporate asset. Management has a responsibility to protect it. Stockholders can claim negligence and hold company executives responsible if this asset is lost due to improper protection efforts. Simple LAG (locks, alarms & guards) will not appear to be proper protection. Further, the law only protects those who protect themselves. You can't just wander into the court room crying "They stole my business secrets," and expect help. You have to show the extraordinary steps you took (and maintained) to elevate your business information to business secret status. Simple LAG will not appear to be extraordinary. You need to inspect regularly. Detecting espionage before the effects surface will keep you out of court, a time and money black hole.

COUNTERESPIONAGE IS NOT A DO-IT-YOURSELF PROJECT

- Don't buy eavesdropping detection gadgets.
- Don't hire a private detective and let them play debugging expert.
- Don't play detective.

Eavesdropping detection is a serious business. Counterespionage work is a full-time specialty within the security field. Professional help is available. Be pro-active, and remember—you may only have one chance to "do it right."*

---

*Kevin D. Murray, CPP, CFE, and BCFE, has been solving electronic eavesdropping, security and counterespionage matters for business and government since 1973. If you have additional questions he may be contacted at *murray@spybusters.com.*

Chapter 13

Power Supply for Security System Devices

Beginning with the explanation in Chapter 7: Electronic Lock Control and continuing up to this point, all of the devices that support security systems require electrical power. Power is needed for lighting, electric door locks, intrusion detection and CCTV systems. Fortunately, compared to the total electrical needs, these systems require only a small amount of power. But the security system power supply needs to be ample, reliable and redundant. If power to the security devices fails, the security system's integrity is compromised. A facility would still have the people and if a facility has chosen to have guards, the guards are still available, but the systems that support the guards, like the computers and communications may not be operating.

This chapter explains how to make sure the facility has a good clean power supply for the security systems.

SINGLE LINE DIAGRAM

Major facilities should have a drawing on file called the single line diagram or the single line power distribution diagram. This drawing or set of drawings shows at a glance how the power supply is distributed through the facility. The single line diagram is more like a sketch than a design drawing. It does not show where things are, although sometimes room numbers are shown. The single line drawing shows the sequence of distribution of power. It shows how power comes into a facility, and where the panel boxes are in comparison to the pieces of equipment. In effect, the power cascades down from the supply through the circuits for all of the various types of equipment to the devices.

Power for a facility is usually divided into four categories. They are: LIGHTING, EQUIPMENT, POWER OUTLETS and ALARM AND SIGNALING systems. The single line diagram shows which circuits are for lighting, which are for equipment, etc. and it shows where the circuit breakers are that control those circuits.

Lighting circuits are just that. They are the panel boxes, breakers and circuits that provide power to light fixtures. Usually these are the normal light fixtures. Emergency lighting is usually powered separately.

Equipment circuits are dedicated power circuits for building systems like the air conditioning fans and water pumps. If a facility has special machines, like milling machines or presses, these circuits are also shown on equipment circuits.

Power outlets are the plugs on the walls. These are what people plug coffee pots and computers into for individual power use. Power outlets can also include special plugs for high power machines.

Alarm and signal system power is dedicated to the fire alarm, the security systems, and the telecommunications equipment.

FIRE ALARM SYSTEMS

While this book has not discussed fire alarm systems, it is important to understand the difference between fire alarm and security electrical power. The fire alarm system is a safety device that is highly regulated by building codes. In fact, the building code for fire alarm systems and the building code for electrical systems are written by the National Fire Protection Association whose headquarters are in Quincy, Massachusetts. In the United States and in other countries with buildings constructed to US Standards, the electrical circuits for fire alarms are required to meet the wiring standards in the electrical and fire alarm codes.

While security alarm circuits are required to be installed according to the National Electrical Code, they do not necessarily have to be installed according to the Fire Alarm Codes. So it is important that fire alarm circuits be kept separate from the security system circuits unless fire protection engineers, and perhaps even the local fire marshals, are consulted. The concern has been that the power for fire alarms is more important in building electrical systems. If fire alarms and security systems share the same circuits, the Fire Code requires that fire alarms and signals have priority over the security signals.

Other developed countries like France, the United Kingdom and Germany have their own electrical codes. Underdeveloped countries do not have as strict a requirement for building code compliance. However, good practice dictates that recognized standards be followed. The United Nations works with underdeveloped countries to help them define for themselves which codes to follow or whether to develop their own. The bottom line is that electrical devices and standards from one country may not be compatible with electrical devices from another country. Hence the facility manager performing electrical work outside the United States and its territories needs assistance from professionals familiar with power supply codes and fire alarm codes in that country.

MAIN POWER SUPPLY

Power for a facility usually starts at a power line near the facility. The power is brought in through cables that are either placed directly in the ground with special wire insulation or they are pulled through conduit buried in the ground. Power cables enter the facility and are routed to where distribution of the power begins.

Another way for power to be fed to a building is through overhead power lines attached to a mast on the building although this type of power supply is less secure than the underground methods.

Overhead power lines are usually bare wire, and are connected to insulated cables before going into the ground. If power is fed underground, the cable or conduit is usually only two or three feet below the surface.

A second way for power to reach the facility underground is for the conductors to be enclosed in pipe. People who are familiar with the trade call the pipe conduit. The pipe can be metal or plastic. Conduit is preferred over direct burial cable because the conduit protects the pipe from digging better than the insulation on direct burial cable.

For large facilities, like factories, processing plants and larger building complexes, which require more power runs for more different types of equipment a facility may have power in duct bank. A duct bank is a series of tubes, usually gray plastic conduits through which the conductors are run. The tubes then have concrete poured around them. The duct bank is the most secure of the power supplies, because not only are the conductors protected by being underground in conduit, but concrete has been placed around the conduit. Duct bank is the most expensive, requiring trenching, conduit and concrete. Also, duct bank can sometimes interfere with other underground piping like sewer and water lines, so installation of duct bank has to be coordinated with these other services. Depending upon the voltage of the power lines, transformers may be necessary to change the voltage from high to low.

Inside Power Cables

Once inside, the power cables are fed to panels where electrical circuit breakers are installed in large gray or green cabinets called power panels. The more power needed, the larger the cabinets. Inside the cabinets, power cables are attached to metal bars called bus bars. Then other circuits are attached to the bus bars for distribution of the power down stream. Depending again upon the voltages, additional transformers may be required to lower the voltages further. From the cabinets, smaller wires are routed through the facility to where power is needed. The size and shape of the cabinets is designed to prevent short circuits and arching. Hence for safety reasons, extreme caution is required if these boxes are opened with the power still supplied inside.

Inside the building, power cables can be carried in either conduit, or another type of wire carrier called cable tray. Cable tray is simply that, a tray hooked together and usually run above the false ceiling. Cable tray is designed to carry the weight of the wire and the bends are designed to protect the wire so it does not crimp or kink. Figure 13-1 shows exposed cable tray in a school.

In some facilities, the cable tray is above the false ceiling.

Building electrical codes have rules about how many wires and what sizes can be put in conduit and cable tray. Also, the codes have rules about what types of wires are allowed in cable tray. Different circuits are not allowed to be mixed because a fire in one type of circuit could spread to another and disable that system.

The single line diagram should show which circuits are kept separate.

EMERGENCY POWER SUPPLIES

For a facility that requires it, a backup power supply can be provided. This is common for many types of facilities like hospitals and police stations. Other facilities can have emergency power but this increases the cost of a facility. The decision to have an emergency power supply is usually a function of economics for those facilities that are not required to have an emergency power supply by the building codes.

Figure 13-1. Cable tray. (*Photo by Robert Reid, 2004*).

Emergency Generator

For facilities that need to have an emergency power supply, the most common practice is to provide an emergency generator to generate electricity if the main power supply should fail. Usually this device is a diesel or gasoline engine hooked to a generator. By starting the engine, the generator produces electricity which can supply the facility with power until the utility power is restored. Depending upon location and need, sometimes two emergency generators are needed. Figure 13-2 shows an emergency generator for a small medical office building.

In general, most emergency generators are diesel generators, because diesel fuel is less hazardous than gasoline. If there is a diesel fuel spill during an emergency it is not as dangerous as gasoline would be in the same instance.

Emergency generators are sometimes located inside the building. If they are they are usually on the main floor. Usually there are large louvers to allow air into the room where the generator is located. This air is required for cooling. The exhaust from the motor is muffled, just like an automobile engine, and piped outside to prevent fumes from entering the building.

On other occasions, the generator is placed outside the building on a concrete slab. Today, most emergency generators that are outside have an enclosure to protect them.

When the main power goes off, the controls of the emergency generator are set in such a way that the emergency generator will start automatically. Depending upon how big the generator is, there are tools to make sure it starts quickly. The industry standard is that the emergency generator must start and assume the loads within 10 seconds of a power failure.

Automatic Transfer Switch

When the diesel engine of an emergency generator starts, it still does not have the ability to carry all of the power loads, so a device called an automatic transfer switch is located between the emergency generators and the distribution panels. In some cases the automatic transfer switch is located within the cabinets of the distribution gear. The automatic transfer switch or switches are shown on the single line diagram.

When the utility power fails, the emergency generator diesel

Figure 13-2. Emergency generator. (*Photo by Robert Reid, 2004*).

motor starts. Once the motor is up to speed the engine generator set begins generating electricity. When the power has stabilized, the automatic transfer switch transfers the power cables to the emergency generator and leaves the power lines to the utility open.

Depending upon the design of the switching system, the power can transfer back to the utility automatically when power is restored, or the system can stay on the emergency generators until someone verifies that the utility is stable, and then transfer the switch so the power is being fed from the utility.

Because of power cycles and other slightly technical problems, it is necessary for the automatic transfer switch to release from the utility BEFORE attaching to the emergency generator. When this happens, normal power is interrupted for a moment. It is sometimes only a flicker, but it is enough that all computers and other electrical/electronic devices stop processing for a moment.

Since most of the other electrical and electronic devices have computer chips in them, they will also stop working. For a security or fire alarm system, this might require that the system be "reset." It means the system needs to be turned off for a moment, and then turned on again.

The same thing is going to happen when the power is restored, because the transfer switch has to switch back to the main power supply and another momentary flicker or power interruption ensues.

Uninterruptible Power Supply (UPS)

In order to prevent this momentary flicker and having to reset the security and fire alarm systems, a third type of power is shown on the single line diagram called the uninterruptible power supply system. Uninterruptible power is electrical power backed up with batteries that provide power in the interim between the loss of the main power and the take up of the emergency power. This emergency essential power is usually a very small amount compared to the rest of the power supply and it is rare for any motors or fans to be on the emergency essential circuits. The batteries for UPS systems are sized to operate the system for a limited time, usually only a few minutes since the emergency generators should provide backup power within 10 seconds. However, if the emergency generator does not start, there needs to be enough emergency essential power to provide electricity in order to allow the building to be evacuated or other measures taken in the event of the loss of all electrical power. Most uninterruptible power supplies have enough electricity for about 30 minutes. This is enough time to either evacuate the facility or to manually start the emergency generator. Some power plants and more complex facilities may have more battery backups, some up to 8 hours.

The fire alarm and signaling systems are required to be supplied by uninterruptible power circuits. Also, the telecommunications equipment and any security systems are sometimes provided with uninterruptible power. Life safety equipment is also on uninterruptible power supply, for example, the power outlets near patient beds in a hospital.

Smaller and less highly regulated facilities can sometimes get by on battery systems alone. Commercial office buildings, for example, may have battery backup for the fire alarm and signaling systems and no emergency generator. If the power goes off for a few minutes people can wait for the power to be restored. If power is off for a long period, these people can be allowed to go home. For a police office, this is not acceptable and emergency power must be provided for long enough to get through the power outage.

Individual computer systems can be supported by stand alone UPS packages that cost from $50 to $150 dollars. They are a battery pack that plugs into the wall and the computer power supply plugs into the battery pack. The batteries remain charged while normal power is supplied and provide the power for the computer and its peripherals when the power goes off. Depending upon price these devices can provide power for from 5 minutes up to 72 hours. Figure 13-3 shows a small UPS system in a television control room.

MAINTENANCE

Emergency power systems require regular testing and maintenance. For emergency systems, it is usually a requirement to test run the emergency generators once a week or once a month to assure all these systems are ready to go in the event of a power failure. A diesel motor needs regular oil changes, the same as an automobile. Electrical switchgear has to be tested periodically by testing automatic transfer switches and other switches. In general, small electrical breakers do not require testing as their operation is certified by the factory.

When emergency power tests take place, it is best to notify those responsible for facility security before the test, as the security personnel would not know whether it is simply a test or a real event.

CONCLUSION

Security system power supply is essential. If the power fails the security system will probably be needed more than at any other time. An intruder or group of intruders may try to compromise the site's security systems by cutting the building power supply. The power is a critical component of the site's security

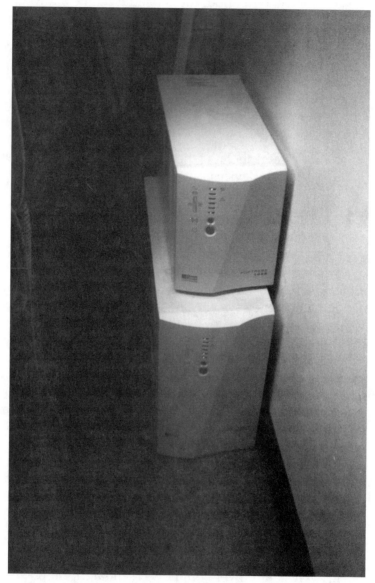

Figure 13-3. UPS system. (*Photo by Robert Reid, 2004*).

system. The next chapters examine the people and how they work with the mechanical and electrical security systems. Guards and guard forces are discussed in the next chapter.

RESOURCES

The National Electrical Code® NFPA 70 published by the National Fire Protection Association, Batterymarch Park, MD 2004.

The guardhouse on one facility was designed as a mini fortress. It was built of reinforced concrete and included gun ports. The emergency generator for the security and communications systems was included. Blast plates were placed over the louvers of the emergency generator openings to prevent the radiator of the generator from being shot out. All of the glass in the facility was bullet proof glass and access was allowed only from inside.

Chapter 14

Guards and Guard Forces

Now that the facility has established access control and the ability to monitor electronically with Intrusion Detection Systems and CCTV, the next step to improve and enhance security is to determine whether or not to utilize a guard force. Guards can maintain vigilance over a facility in support of electronic systems or they can provide a stand alone operation. Guards also cover events, activities and provide escort services for personnel who need protecting. Many times guards are brought in to perform surveillance over a facility while electronic systems are being installed or when those systems have been temporarily shut down for maintenance. There are several options open for a facility as they evaluate the need for a security force.

NUMBER AND FUNCTION OF GUARDS

With electronic detection and CCTV systems, the number of guards required is greatly reduced. These devices act as the eyes and ears of a guard force to tell them where and when there is an intruder. For retail applications, a security guard represents authority and is a very strong deterrent to shoplifting and employee theft. The presence of guards may also deter violence. One of the first determinations to be made when deciding if security guards are needed is to figure out how many guards are necessary.

In order to have an accurate assessment of the number of guards required it is necessary to determine the functions the guard will need to perform. The security guards can cover the facility around the clock, twenty-four hours a day, seven days a week, or they can cover the facility at night and on weekends. Some facilities only utilize guards during the day when the facility is open. Others only need guards at night and on weekends when the facility is closed.

237

Seven Days Per Week, Twenty-four Hours Per Day

For guard coverage around the clock, it requires approximately 4 men. There are 168 hours per week, but laws govern the amount of hours worked in a week to about 40. More hours can mean more overtime. In many facilities the shift is 8-1/2 hours, the extra 1/2 hour being a break for the guard for lunch. This is enough overlap for 4 guards to cover one position in a facility. Some facilities are able to cover with 3 persons. Each person works 8 hours, which totals 24 hours in a day. But in this case each person works every day. There is no time off for weekends or personal time. Covering the facility this way works for a short period, but extra overtime is required since each person in this rotation works 56 hours in a week. In this scenario, sixteen hours of the week are premium time and require extra wages.

Another short term solution for around the clock operation is for two persons each to work a 12-hour shift. In this scenario, each person works 84 hours in a week, giving each person 44 hours of premium time. Since this premium time is costly, it is less expensive to have three or four persons working than it is to have two persons working a 12 hours on and 12 hours off schedule.

Care must be taken to rotate the shifts. Usually, nights pay a premium of 10% above base and Sunday may carry a time and one-half or even a double time premium. Utilizing the 4 man rotation requires each person to move from shift to shift each week, which can be physically hard on employees.

Knowing what hours the facility is to be covered by the guards is the first step, and knowing how many persons it takes to cover the hours is the second step. The third step is to determine how many guards are needed to cover the facility. This number is a function of many factors which include hours the facility is open or closed, size of the facility, and the number of entrances and exits to be covered. The door access control system and the CCTV system make this coverage much easier because it can record and tape the comings and goings of persons any time of the day or night.

For an office building that is guarded at night, one guard without CCTV can patrol the facility and check the exit doors and look for intruders. For a larger facility, with higher security standards it may be necessary for two guards, one to patrol and another to guard the entrance. It may be that the guard patrols the facility once or twice per shift, checking entrances and exits. If all is secure, they may not have another function.

For a multiple building facility, depending upon size and number of entrances and exits, more guards may be needed. On some facilities, the guard force includes up to 40 persons. This includes uniformed armed guards covering the entrances to the facility and to individual buildings, roving patrols and an emergency team that is held in reserve to cover any intrusion events for 24 hours per day, 7 days per week.

On the reduced side, perhaps all that is needed is for a guard to check the facility once in the evening to make sure the doors and gates are locked.

UNIFORMS

Most security guards wear uniforms. And the normal color is either blue or blue with white. This is because, historically, the color of police uniforms has been blue, dark blue, navy blue or black. See Figure 14-1, Typical Physical Security Guard Uniform. Some countries use white as a color for law enforcement and a security guard's uniform may be a combination of white and blue. Historically, police uniforms have had brass buttons which offset the dark blue and many uniforms today have the a patch sewn onto the shoulder and over the left breast that is bright gold, the same color as the old brass buttons on original police uniforms. A uniform indicates that the person in the uniform is "united" with a group dedicated to the same thing. The law enforcement uniform has become a symbol of authority in many countries of the world. Since the guard force represents authority, members of the guard force may also wear a uniform. The uniform shows all visitors and guests that the person has given up his individual clothing in order to unify with a group. In the case of a security guard the uniform says that the person is present to maintain order and protect property.

The uniform shows that the individual is part of an organization of more than one person. And it shows that the facility is committed to security by providing a person to perform the task of guarding or protecting the assets.

Uniformed security guards are often seen at public functions, retail stores, and military bases. The uniform tells non-uniformed persons that the person in uniform is committed to asset protection and that these persons are under certain rules and responsibilities of law and order.

For outside work, the uniform should include a hat or cap, for inside work, a cap is not necessary.

For a few facilities, the choice is made to forgo the use of uniforms, but even in these facilities, the security forces will dress appropriately, perhaps in a jacket with a white shirt and necktie. It is important that the security guard be distinguishable from the common public.

Depending upon the state, separate licenses may be required for certain uniformed activities like private security guards.

Figure 14-1. Typical physical security guard uniform. Source: Public Domain Photo.

FIREARMS

Depending upon the facility's needs and expectations, the decision for security guards may include the carrying of firearms. In most cases, this will mean carrying a pistol or a handgun. There are very strict rules about armed guards who carry firearms. In general both the person and the weapons are required to be registered with the state and local police. The registration requires that a security guard who carries a firearm has been trained and certified to use the weapon under conditions that most managers never expect to encounter. The certificate to carry a firearm also includes a background check which is conducted by the state agency. For additional information on background checks for employees see Chapter 15: People Systems.

For higher levels of security, guards may operate in teams of two, three or more. They may be trained to use more elaborate weapons. Usually, the next weapon for an armed security guard is a shotgun. More elaborate weapons, such as automatic weapons are usually a matter for local law enforcement and will not be carried by private security guards.

Depending upon the threat, armed guards may also be required to wear bullet proof vests. This is obviously to protect the person not only from assailants with firearms, but from knives and clubs.

For a facility where weapons are carried, there are additional criteria relative to the weapons themselves. For example, lockers should be provided for holding the weapons when the guards are off duty. Some persons may prefer to carry their weapons with them at all times, but if the facility provides the weapons or requires the carrying of a weapon, then the facility may wish to keep the weapons rather than allow the guard staff to carry the weapons home. In addition to weapons, consideration must be made for ammunition. For obvious safety reasons, ammunition must be kept as safe as the weapons.

In addition special requirements for ammunition, training is required for guards who carry firearms. Most guards who carry firearms are required to have been trained in the use of the weapons they carry. Usually, a state or municipality is responsible for issuing a firearm permit that allows the individual to carry a weapon on the job. For training the guards can return to the certifying facility to maintain their training. If the facility is remote, and the area will allow it, a target shooting range may be provided for the armed guards to maintain proficiency.

For a facility that needs to utilize armed private security guards, close coordination with local and state law enforcement is required and extensive legal requirements must be met. Usually, local law enforcement will recommend to a facility when conditions warrant the hiring of armed private security guards. And often, local firms that provide security guards are familiar with the law enforcement requirements.

Obviously for a facility that chooses to provide armed guards, the policies and procedures are extensive and must be met. A facility that has armed guards also needs to coordinate their liability and risk with their insurance providers.

VEHICLES

For various campuses the facility may choose to provide the security guards with a vehicle for patrolling the grounds. The vehicle, as with the uniform, should be clearly marked to show that it is a patrol car and not simply another vehicle. For the same reason as the uniform, in an event, law enforcement needs to be able to identify the security guards and differentiate them from intruders. Another advantage of the clearly marked patrol vehicle is that it will reduce the tendency of passers by to call the police when they see a vehicle patrolling the grounds.

GUARDHOUSES

For physical security guards who control access at a gate, it is common to provide the guards with shelter in the event of inclement weather. This usually takes the form of a guard house which is discussed in Chapter 3, Physical Separation. The guard house may require toilet facilities. If not, some method of relief for the guard is required for reasonable work hours. Guardhouse guards may stop vehicles, search them, inspect employee and visitor badges and inspect cargo. In some facilities guardhouse guards also issue temporary badges, but this can become a complex process if the decision is for the guards to issue temporary badges at the facility gate, because all other vehicles are held up while the guard issues a badge to the visitor. There is more information on badges in Chapter 15, People Systems.

For this reason, many facilities have additional guards on day shift for issuing badges. One guard maintains the gate and another takes candidates inside for issuing badges. Depending upon the level of security, some facilities allow the vehicle to pass into the facility, park and come into the guard house for badging. For higher security facilities, a person without proper credentials is turned back and the guard force relies upon others to issue badges or provide necessary credentials. This can be frustrating if a person's credentials are found lacking on night shift or weekends. Many managers have shown up on a Monday with a group of angry customers who were locked out of the facility over the weekend because there was no one to issue a badge and the guard force kept them out all weekend.

COMMUNICATIONS

Radios

As the decisions are made about the size and makeup of a security guard force, an important consideration is the method of communications for the guard force staff. Most guards carry two way radios that enable communication with other members of the force. Depending upon the level of security the radio frequency for the guard communications can be important. A sophisticated intruder may try to jam the signals of the guard force radios to assist in delaying the guards. Usually guard radios have multiple channels allowing different types of communications on different channels. Radios have become quite sophisticated and can be carried on the hip with a headset for hands free communications or the radio can be carried with a hand mike that is clipped to the shoulder. Finally the radio can be hand carried. The radio has to be "keyed" in order to talk and once a mike is "keyed" all receivers within the range of that channel will receive that signal.

If two people key a mike at the same time, the signals cancel each other. This term is called "walking on." If one person keys a radio mike and before he finished speaking, a second person on the same frequency also keys a mike, it is said that the second person has "walked on" the communication of the first person. Only radio discipline can prevent this type of problem. This is why radio communications are often very short and a lot of abbreviations are used. In some cases, a response is a simple click of the mike. More information about emergency communication is included in the chapter on Emergency Response, Chapter 16.

Telephones

In addition to radios, communications with guards and between guards can be conducted by telephone. Depending upon the facility, there may be telephones within the complex from which a guard can check in while making rounds.

Cellular Telephones

Many private security guard forces now provide the guards with cellular telephones that allow the guard to communicate with their counterparts and with clients or customers. If a guard making rounds discovers a gate unlocked, the guard can telephone the owner or another responsible person and determine whether to secure the gate or leave it unsecured at the owner's request. Since owners do not often carry radios, the cellular telephone can be used to call the owner to inform him or her of important events on the property. The advantage of the cellular phone is that the guard can tend the gate until a determination of what to do about the unlocked gate is made. Otherwise he may have to leave the gate untended until he can find a telephone.

Computers

Finally, guard communication can be via computer. A guard can occupy a workstation and communications can be by email or other computer tools. For guards who make rounds or patrol, this option is not as feasible. Plus, since nights are long, the guard is going to be tempted to utilize a computer for purposes other than work which may be unacceptable.

ROUNDS

For most facilities that have physical security guards, the guards make rounds of the facility on a routine basis to check openings, vehicles, areas or spaces. For many facilities with electronic access control as discussed in Chapter 9, the guard's use of his proxy or magnetic badge can be used to track the route of the guard through the facility. Other facilities utilize special devices with electronic signature keys to track the guard's progress during the shift.

LOGBOOKS

Again depending upon the facility, the owner may choose for the guards to record their activities in logbooks. These logbooks are usually written up at the end of the shift but can also be used throughout the shift if necessary. Logbooks have an advantage over computer type logs in that they cannot be altered, although computer tools can also be created to protect log book entries into a computer database. Hand written logbooks usually are bound, have high quality paper so that the ink will not smear and the page numbers are serialized. That is, every page is sequentially numbered to prevent altering the log by removing pages. The process of making logbook entries is a matter of facility choice and, if this is a choice, the facility manager must recognize that each individual's logbook writing will differ from others. Some entries will be terse one or two sentence affairs; others will be long flowing narratives. Logbooks should be kept in indelible ink because the purpose of the logbook is to maintain a permanent record of events. Times and locations should be entered. For a sophisticated facility, these times can usually be verified with event control logs from the door openings and closings from the computer software. Some guards may not write well. Because of this they will not be comfortable logging the events and, as a result, their log reports tend to sketchy or spotty, there will be misspelled words and other minor grammatical problems.

Guards should not have collateral duties. They are guards.

PAY

The pay of guards varies by area. Major factors in guard payment are the same as for many other facilities. It is possible to hire guards for minimum wage, but these guards will be unlikely to have had a background check and will have little or no training. This is not recommended for a facility but it may be acceptable for an event. Unfortunately, a guard or guard force is there to protect against something the facility wished would not happen. Therefore the temptation will be to pay the guards a minimum amount.

According to a recent survey, the median expected salary for a typical Patrol Officer in the United States is $42,800. The range is from $35,656 to $50,482. This rate is for an employee with benefits including health insurance and vacation pay. Guard forces, in general, will make less although a facility with high security standards may decide to pay the guards more.

For guards with extra skills, the price increases. For example guards may be needed who have multi-language skills and of course guards who carry firearms are better trained with more credentials. Therefore, armed guards are more expensive than guards who do not carry firearms.

Finally, a union may govern the use of guards in an area or locality. In this case union agreements will require compliance with pay, hours of work, overtime pay and many other cost factors.

SUBCONTRACTING
FOR PRIVATE GUARDS

Because of complications involved with an in house guard force, many facilities choose to contract out the function of guarding the facility to a local, regional or national security firm that specializes in providing security guards. There is some advantage to contracting out since the facility may be able to limit its liability by separating the responsibility between the facility and the company that provides the guards. The subcontracted firm provides armed or unarmed guards, communications equipment and, if necessary, patrol cars. In addition, the contractor performs background checks of the employees, provides the necessary certifications, radios, and policies for what to do in events.

In general, guards cannot arrest people, and policy is that the security guards will notify police if a crime is committed. The main purpose of the guard is to observe and report and to be called as a witness if a crime is committed. The guards can control access by maintaining gates and doors closed and locked. Guards, if properly trained, can respond to facility alarms and assist law enforcement in determining if the facility has an intruder.

Most contracted security firms use a standard contract. When the facility requires assistance, officials of the security firm will expect to visit and tour the facility. What they do is, in effect, conduct a miniature risk assessment and determine how many guards are needed, what their function is, and the level of expertise necessary for the guards. When this is completed, the security firm will make a recommendation as to the scope recommended for the facility. If this is not acceptable to the client, the firm may agree to cover the facility with a disclaimer in place. This means that the facility has chosen a level of security protection at less than the amount recommended by the security firm.

Much of the security firm's contract is devoted to liability. The security firm should agree that the guards are the responsibility of the security company, not the facility being guarded. The security company may insert a clause to prevent the client firm from hiring the guards directly and thereby prohibit the client from "stealing" the guards from the security firm.

Usually, while the guard force is acting as a deterrent, there is no guarantee that the facility will not be subjected to an event. Most security firms will advise and guard, but they aren't responsible should an event like a robbery or theft occur. The client firm should make sure that there is no liability to the client firm should one of the guards be injured on the client's property. Such liability should be the responsibility of the security company.

The contract between the security firm and the client company should include standard contract clauses for payment, bonds, insurance, termination of the agreement, etc.

RESOURCES

Service Employees International Union represents over 20,000 security guards. www.seiu.org.

Chapter 15

People:

Personnel, Badges, Background Checks, Training, System Testing, and Security Professional Certification

Most of the information presented in this book has been about mechanical, electrical and physical systems that are available to help a facility protect its assets. However, since people are one of facilities most valuable assets, it is important to understand what facility measures can be taken to protect the people and to protect assets from malicious employees. What steps can be taken to help a facility protect itself in ways other than hiring guards and purchasing new hardware?

The devices explained in Chapters 3 through 13 are not intelligent. They were devices only capable of limited actions, whether it be indicating motion in the case of a motion detector, or recording the movements of person at the cash register in the case of a CCTV System. It is the people, and what they do, that ultimately determines the strength of a security system. If the alarms go off, and there is no plan, then what good was the system?

This chapter is about the next step in physical security: the people. There are several elements of how people relate to security and this chapter examines these main interfaces. This chapter is broken down into four parts. The first part explains staff issues as they relate to security. How do employees react to security or security system change? Will they be patient while the systems are installed and what about after they are installed? Will they use the systems or will they try to defeat them because they are a nuisance? Does the facility need to badge its employees? How long does it take to badge employees and how much does it cost?

Next, the chapter reviews how to handle problem employees. What about employees that are a threat because they seek to compromise the assets? What steps can a facility can take to protect itself from these problem employees? Training of staff in new security procedures and drug testing of employees is covered.

A section is included about security professionals. What types of training should they have or what certifications can be provided to a facility to assure the security professional has the necessary skills and credentials to do a good job?

Last is a short section on system acceptance testing and maintenance.

STAFF ISSUES

How does the staff feel about security? Do they see the need for physical security as a nuisance? What is the employee reaction to proxy badges, keypad combinations, intrusion detection equipment and key control?

Security Policies and Procedures

Depending upon a facility's current status, a facility may have existing policies and procedures in place for security or there may be no policies and new security systems may require that management issue new policies for the staff to follow. Many of the systems explained in this book would require new policies or procedures. Training is helpful in what the new systems are, what they do, what advantages they provide and why it is necessary to implement them.

The Security Professional as a Salesman

In order for security to be effective, it helps for the security staff to be helpful and patient with regular staff. Remember their mission isn't to be secure; it is to deliver the facility's product or service. Too much security can grind a facility to a halt and this is not good for business. Therefore as long as the security staff can be helpful and assist with the processes while protecting the assets, security will be welcomed. One of the major lessons learned when a security system is modified is giving the staff time to understand the coming changes and providing time for the staff to adjust to the new systems or policies.

It is a natural tendency for staff to resist changes, so patience and understanding have proven helpful as new security systems develop.

Security by its very nature is restrictive. Door locks and intrusion detection devices slow movement so it is helpful if the facility can sell the security program and get the staff to share in some of the decisions so that they will participate willingly as new security systems and programs develop. A facility may have operated a certain way for many years and when the risk assessment reveals a need for a change, the new policies can be viewed adversely by the staff.

Attempting to force an issue can lead to stronger staff resistance and this is not what is needed. What a facility needs is security that protects the assets. As long as the staff understands the value of security to them as individuals they will be willing to support any security measures needed.

There will be staff who will oppose new procedures. This resistance can take many forms from constantly forgetting one's badge, to "accidentally" leaving doors unlocked, to "testing" the system by trying doors which they are not authorized to enter and even to defeating some systems by propping doors open so they cannot lock. While these cases are rare, management needs to think through the policy for these resistant types of actions in advance. It may be necessary to discipline staff members or even to let people go if they do not follow policy. However, the security professional needs to exercise authority with caution because discipline can lead to rebellion and in the end the security professional trying to make the changes ends up leaving instead.

For the security professional to smile when explaining new or current policy to new employees works far better than the scowl and threats to people for not following the policy. Discipline is not easy, especially if there has been little or no discipline in the past. For a security professional who comes into the organization from law enforcement or military, discipline is almost automatic. But for others who are not used to this level of rigor, it is a more difficult challenge to develop the discipline necessary for good security.

Leadership Principles

The following management techniques are common management techniques but are worth repeating for facility and security managers because these tips will help to make leading security changes easier and more effective.*

1. Set an example. A good facility or security manager leads by doing what he expects employees to do. If there are policies, the manager should follow the policies just as they expect the employees to do.

2. Become interested in the staff. Good leaders take an interest in what their employees do and are trying to do. They care about the people who work with them and they let it show.

3. Listen to the people you work with. In order to be credible it is important to establish communications and the best tool for communicating is listening carefully to staff issues and concerns.

4. Gain the cooperation of others. Avoid arguments and show respect for others. Be friendly and smile often. This opens the door to being able to lead. Sometimes it helps to dramatize ideas or to challenge others to do their best.

5. Lead by doing. Appreciate others, and ask questions instead of giving orders. Encourage others. Praise their work and their support of security.

Changing a staff's attitude about security isn't easy. It requires patience, hard work, integrity and diligence.

BADGING EMPLOYEES

The decision of a facility to have employees wear and display badges is one way to upgrade security in a facility. The de-

---

*These 5 principles were taken from the Dale Carnegie Management Training Course©.

cision to badge employees may be a matter of company or corporate policy or it may be the result of risks associated with the risk assessment. Facilities that choose to badge their employees include hospitals, medical facilities, airports, government offices and large corporations.

Offices with secure information and few visitors would be a good choice for issuing badges to identify employees. The primary advantage of a badge is to quickly identify whether a person is authorized access to an area. For a small office where the staff knows each other, a badge for every employee should not be necessary. But, in a large facility with multiple areas, varying levels of security, and occasional visitors, badges are an important security tool to aid in identifying persons who have authorized access to the facility's resources.

Badges, especially photo badges, can also act as a deterrent to unsophisticated intruders and the general public. Badges will not deter a determined or sophisticated intruder. A determined intruder will attempt to crash through the barriers. The advantage of badges in this case is very small. A sophisticated intruder may choose to duplicate the badge and pretend to have authorized access. For this reason the badge policy should include requirements that employees keep the badges out of sight when they are outside of the facility. An intruder could steal an employee badge left in someone's automobile, for example. If the badge includes proxy access cards, the advantages of badging employees and electronic door locks is lost if the badge and proxy badges are stolen.

Badge Sizes

Badges come in several sizes but one of the most common ones is the same size as a credit card (See Table 15-1 Common Badge Sizes). Others are wider and larger and will not fit easily in a wallet or billfold. Badges also come in varying thicknesses; the thicker badges are physically tougher and usually last longer than thinner paper badges printed on cardstock. Some badges come blank from the vendor with a proxy card antenna or smart card chip embedded in them. This would combine the badge with the access codes to open doors.

Table 15-1. Typical Access Control Badge Sizes

Badge Type	Height (Inches)	Width (Inches)	Thickness (Inches)
CR60	2.375	3.25	0.03
CR80 (Credit Card)	2.123	3.375	0.03
CR90	3.63	2.37	0.03
CR100	3.88	2.63	0.03

Badges are often designed with corporate logos, colors and can include an employee photo showing clearly who is carrying the badge. Figure 15-1 is typical of a CR 80 photo badge.

Most badge policies require that the badge be prominently displayed on the front upper part of the body. This policy works well to indicate when a person does not have a badge.

Figure 15-1. Photo badge. Source: Reprinted with permission of the Data Card Corporation, Copyright 2002, Data Card Corporation. All Rights Reserved.

Badge Challenge

A badge challenge is a request to see a badge. However, when an employee conducts a badge challenge, he should be prepared for the unexpected. Usually, the response to a badge challenge is, "Oh, I forgot it." To which the employee who has forgotten the badge should either return for their badge or request to be issued a temporary one for the day. However, if the person is a determined intruder, the response to a badge challenge may be quite different. For this reason, employees should be trained in the policy for conducting a badge challenge. The employee responsibility in a badge challenge should be to refrain from entering a confrontation. Confronting an intruder should be handled by guards or law enforcement. If the response to a badge challenge is unfavorable, the employee conducting the challenge should not attempt to forestall or detain the intruder, the employee should notify appropriate personnel of a possible intrusion. Any attempt to detain or restrain the intruder could lead to violence which is not what the average employee is prepared to handle. Issues of emergency response are explained the next chapter.

For a large facility, with more than 250 employees, facilities sometimes choose to color code their badges. Color coding the badges by varying the colors or adding stripes or different backgrounds on the face of the badge is used to indicate the area where the employees are authorized. By using different colors, employees who know the colors can recognize employees who are in unauthorized areas. The colored badge indicates an employee is allowed to go into some areas; a different color badge indicates access to a more secure area. Another area may choose third color. These colors are known by employees and guards. A badge that is not the correct color, indicates the employee is in an unauthorized area and may be an intruder.

Vendor Supplied Badging Systems

Many vendors sell a complete employee badging package, which includes a digital camera, blank cards (including proxy cards or magnetic cards) and a special printer to print photographs on the card with indelible ink so it won't rub off. Well built badges will last several years. Some vendor types have badges that include a proximity badge or a smart badge.

Other facilities issue the color badge with the photograph and the proxy badge separately. Then both badges are placed in a clear plastic packet. The identification badge is in the front of the packet and the proxy badge is placed in the back. The plastic packet holds both badges and is either clipped to the employee's shirt or it is clipped to a lanyard that employees wear around their neck. The badge can be unclipped and passed close to the proxy card reader for access. For more information on proximity cards, see Chapter 8, Electronic Lock Control.

The vendor supplied badging systems are capable of badging between 45 and 50 people per hour. So for a facility with 500 employees, badging the staff will take a couple of days. The complete badging system includes the badges, a digital camera, a special printer to print employee identification, colors and photos on the badge, and computer software to collect the employee's badge information into a database. These packages retail for between $300 and $2000 dollars depending upon the quality of the badges and the quality of the employee photos.

One vendor sells special temporary badges that expire. This badge system uses a special ink or paper, which will change color after a day or two. When a visitor comes to the facility and they are issued a temporary badge, the badge becomes striped after the time of their visit expires. The striped badge indicates to the staff that the time for the person's access has expired.

Badge Policy

When a decision is made to badge employees who have not been badged before, the first step is to issue a memorandum to the employees telling them of the new policy requiring badges. The memo should state the reason for issuing badges (needed to improve facility security). The memo should indicate when the new badges will be issued and what the requirements are going to be. The notice should inform employees what information will be collected for issuing the badge, when employees are expected to wear their badge, what is expected of employees concerning the badge when they are not at work and what the policy is for returning the badge when employment ends. The policy should state whether the employees are expected to take their badges home or whether they are to leave them at the office. The issue of badge challenge should also be addressed.

Gathering of employee information for badging may be subject to the privacy act so human relations and legal counsel should review the policy before it is issued to the employees.

Badging Process

People can be badged as groups, by team or work unit, or they can be badged in alphabetical order, with people whose names begin with the letter A coming in first thing in the morning, people with names beginning later in the alphabet being spread out through the day. The badging process should be as simple as possible and the facility should provide the lanyards and plastic pockets as well as the badge in order to encourage uniformity.

Setting up the badging process takes about a week. This includes the time to set up the equipment and test it. The staff who will be preparing the badges should have time to become familiar with the equipment. Depending upon the number of employees the process of issuing the badges takes a couple of days. Plenty of spare badges and ink cartridges for the printers should be available. If the badging equipment breaks down, the process will have to be halted and rescheduled. Any malformed or misprinted badges should be retained and destroyed by the crew making the badges.

If badging employees is tied to a new door locking system, is common to issue the badges a week before turning on the door locking systems to let the staff get used to carrying the badges before having to use them. Remember it is a new process for employees and they need time to become used to regularly carrying the badges and using the new proximity door locks.

Replacement Badges

After the badging program is established and employees are regularly carrying their badges and using them, the occasional new employee can be badged when he or she joins the organization. In many organizations, photos are taken only on one day of the week for new employees, usually a Tuesday, which gives the new employee time to show up on Monday and sign in and meet the staff. Some organizations badge the employee as the first order of business. A good organization usually has collected the employee information for the badge prior to employment so that all that is necessary to make the badge is the photograph. As a courtesy to new employees, management or staff may wish to warn the new employee that they will be photographed for their new badge.

Of course photographs for badges should not allow wearing of hats, dark glasses, or hair that covers the face. Beards and mustaches should be allowed. For some religious groups, beards and hair are sacred, so any policy for employee photos should allow for facial hair.

When one type of badge has been in use for a long period of time, when some badges are worn and others have been lost and when employee turnover has changed the makeup of the staff, it may be time to reissue badges. Reissuing badges for an entire facility is similar to initial badging, however since many of the employees are still in place, much of the information is the same. Given that badging equipment is largely computer driven, it may be faster and more economical to procure new badge making equipment, rather than try to reuse the old system.

Finally, badging employees only works in some types of installations. It is difficult, for example, for construction workers who are climbing, laying down, crawling under or over equipment to maintain badges in the same manner expected of office workers who dress differently and usually work at a desk. In a work environment like a construction zone, badges get lost more often. Of course, wearing a lanyard around any type of rotating machinery can be unsafe and would not be recommended.

Occasionally, employees resist carrying a badge because it is considered an intrusion into their privacy, or it interferes with work in some way. Problem employees can lead to security problems, which is the next topic in dealing with personnel matters.

STAFF PROBLEMS

For many organizations the greatest threat is theft from within the organization. This can be the result of many elements but when an employee has access to the facility it is much easier for this type of person to compromise the assets than for an outside intruder. In the retail industry, far more theft is the result of employees taking goods than from shoplifters. Other employee problems can be the result of stress, drug or alcohol abuse, financial difficulty or crime. Workplace violence is also a potential problem which can be triggered by any number of emotional or social factors.

For this reason, most facilities have decided to conduct background checks on employees to determine past history of crime or acts which could lead to problems. Another method of checking employees is a test for illegal or illicit drugs.

BACKGROUND CHECKS

Previous history is the most logical and effective way to determine whether an employee is prone to problems. For this reason many facilities elect to perform a background check on new employees. New companies may make it a matter of policy to initially check the background of all employees. A background check is usually a sensitive matter to employees and, since the facility will be checking into a person's history, some resistance may be encountered.

Some corporations contract out the process of the background check because of the sensitive nature of the checking. All information collected on an individual is subject to the privacy act and, consequently, information must be carefully controlled. Rumors of illegal past activity about one employee can damage a reputation and lead to a lawsuit. By contracting out the background check, the employee's recourse is on the contracted firm which can limit a facility's liability a little bit.

Background checking starts with asking the employee a few basic questions and then verifying that information. Depending upon the sensitivity of the information in the employee's position, more detailed information can be obtained. However, it is important to make sure that the information that is collected is the same for all employees in that job category. Also, since background checking is a sensitive subject, the questions asked should be screened by the firm's legal counsel to make sure the questions being asked do not violate any local laws.

Depending upon the sensitivity of the position and the level of responsibility, a background check can be simple or complex. A simple check might be a credit history, a criminal records check or driver's license check. A complex background check would include these elements but could go into the person's background more thoroughly.

The United States government performs a "National Agency Background Check, (NAC)" on sensitive positions within the government. The National Agency Background Check includes an FBI records check, and a check of information in other federal and local criminal agencies' records. The form for the National Agency Background Check is the Untied State's Office of Personnel Management Standard Form 86. It can be found on the World Wide Web at http://www.opm.gov/forms/html/sf.asp. The government's background check is very elaborate and correspondingly expensive. It is authorized by law for a few selected federal employment positions. Federal contractors in sensitive jobs are also required to undergo the federal government's National Agency Background Check. It is also the first step in the more sophisticated Federal Secret and Top Secret security clearance programs.

A background check usually includes a check of the following information:

Personal Information

Is the person truthful? The first element of a background check is to verify if the information presented by the candidate is true. This can be performed a number of ways from checking a person's driver's license information or another form of personal identification. (For people who do not drive a car, the Department of Motor Vehicles will issue a personal identification card that is not a driver's license.) Also, telephone books can be reviewed to verify if the person lives at the address claimed. The telephone number where the person can be reached can be called to verify that the number is the phone number of the person. For a person who is not truthful about their name and where they live, this can be an indication that an candidate is not a good risk.

Litigation History

Court records can be checked to determine if the candidate

has been sued or subject to other legal, but non-criminal activity. This information is a matter of public record and can be obtained from the local or county court where the person has lived. A person may have been sued in civil court for non-payment of debt or marital problems. A divorce is a matter of public record. The substances of civil liability background serve to indicate whether the person is reliable for work or to pay debts.

Criminal History

As with the records of civil liability, criminal records can be checked. These documents are also a matter of public record. Depending upon the results of this records check, a history of theft or violence could emerge.

Bankruptcy Check

As with the civil records check, any history of bankruptcy can be verified. This type of information may indicate that a person cannot pay their debts.

Credit Check

A credit check can be conducted to determine whether a person regularly pays their debts or is in arrears. A person with large debts may be on the verge of bankruptcy.

More thorough investigations can be conducted from visiting the person's neighborhood and talking with neighbors, which is one thing that the Government does when conducting its National Agency Background Check. Also, former employers can be contacted and professional agencies and licensing boards will have records that are in the public domain.

The person may be asked to release information from medical records. Communication with former medical providers can indicate health problems, or psychological or emotional problems. The release of medical records on an individual without the individual's permission is considered an invasion of that person's privacy.

Regardless of the information collected, the results of the background check are only used to screen potential problems. Just because a person has had problems in the past is no reason to deny them opportunities in the future, but a past history of problems can indicate a potential for future problems. Facilities need to exercise caution with the results of background information. Denial of employment can lead to legal problems. So, before conducting background checks it helps to utilize the resources of legal council to make sure laws are not violated and liability is limited.

Most background checks go back 7 years. And checking is conducted for every place the candidate has lived in that period. Some sophisticated background checks go back as far as there are records. In addition to information on the candidate, the Federal National Agency Background Check verifies the place of birth of each parent and the location of each sibling. Any birthplace outside of the United States leads to further investigation and records at that location. Investigators also talk with spouses, former spouses, family members or former co-workers. If the investigator finds issues, they are presented in the candidate's file for the decision makers. More information is presented on setting up a security program later in this chapter.

For the less sophisticated background check, a number of web sites will perform a background check for a nominal cost of $25 to $100. These web service providers conduct credit and criminal checks, and personal history information checks. A typical web site for background checking is located at www.backgroundcheckgateway.com. This site offers to perform background checks electronically through public records on the World Wide Web. The price for the results of these searches varies from $7.95 for a simple address check to $59.95 for address, criminal records and public property records check. For more, you can even have a personal records assistant help with the information needed.

DRUG TESTING

In addition to conducting a background check on an individual, many organizations conduct pre-employment drug testing. Federal statistics indicate that approximately 7% of individuals over the age of 12 used an illicit or illegal drug in a typical month in 2001. For persons in the 18 to 25 year old category the rate of drug use was 18.1%. Of this number, almost 76% are either full time or part time employees. The rate of illegal drug use is higher in metropolitan areas. Approximately 48.3 percent of people over 12 use alcohol and nearly 10% reported driving under the influence of alcohol at least once in 2001.*

Obviously, there are a number of reasons for drug testing, but fundamentally, a drug test limits the liability of the company by determining if the potential employee is a threat to others and to themselves because they may be drugged on the job. Drug testing has come a long way since it became legal to drug test employees in the late 1980s. Laws in different states vary as to what types and numbers of drug tests are allowed so a legal review is a requirement before setting up a drug testing program. Many court battles eventually validated the employer's right to drug test employees provided they have a written drug testing policy.

Employers who provide services in the transportation industry, i.e. truck drivers, airline pilots and locomotive engineers are required by the United States Department of Transportation to conduct drug tests on their employees. Statistically, people who use illegal drugs are more likely to become involved in an accident on the job, are 1/3 less productive and file more and higher workers' compensation claims that workers who do not use drugs.

Drug testing is normally conducted prior to hiring and sometimes may be conducted after a theft or an accident. Usually a person is directed to report to a medical laboratory and provide a urine sample. The temperature of the sample is measured to confirm its validity. Some persons have been known to substitute another person's urine in an attempt to fool the tester. This urine will not be at body temperature when provided. The sample is divided into two parts; the first is analyzed by a fast, inexpensive method called an immunoassay test. If the result is negative, then the lab reports a negative result. If this first test is positive, the second half of the sample is analyzed in a machine called a GC/MS analyzer. The GC/MS stands for gas chromatograph/mass

---

*Office of National Drug Control Policy, Drug Data Summary, March 2003

spectrometry. The GC/MS instrument is very accurate and results from the instrument have withstood many legal challenges.

The instruments are set to scan for illegal drugs including:

- Marijuana
- Amphetamines
- Opiates
- Cocaine
- Phencyclidine (PCPs)

The lab can be directed to test for other drugs and can also test for alcohol. On occasion the lab identifies diabetes or pregnancy which can also be determined from urine sample testing.

Facility Conducts Drug Tests and Finds Pregnant Employee.

In the late 1980s when a facility began conducting drug tests the testing company determined that one of the female employees was pregnant even though they were not testing for pregnancy at the time. Because the job was hazardous to pregnant workers and there was a risk that the chemicals in use were dangerous to developing fetuses, the company decided to terminate the female's employment rather than wait to see if problems developed during pregnancy or at birth. The female employee sued the company for an illegal termination and won a large settlement because the written policy did not include testing for pregnancy and the determination of pregnancy by the lab was an invasion of her privacy.

The GC/MS test can indicate drugs are present in a candidate. As a result, the final step in a drug testing program is to call the candidate in and ask them if they are taking any prescription medications that might affect the results. At this point the candidate can present evidence to indicate that the drug screen's positive response is the result of legally prescribed prescription medications. In the event that the prescriptions are presented and are prescribed by a licensed physician, the results of the drug screen are negative.

Since the late 1990's, drug testing has become relatively economical, with drug tests costing less than $500 per test. Costs for legal council to write the drug test policy ranges from $1500 to $3000 depending upon the state where the policy is written.

For many facilities when they find a candidate has tested positive for drugs on a pre-employment drug test, the decision is to reject them as an employee. There is no reason given, just that the employee does not meet the requirements. For employees who have been with the company for a reasonable period, terminating them for drug use would likely lead to a lawsuit. Most companies try to work with an employee who is addicted to drugs by providing counseling and therapy. Drug addiction is a disease and terminating an employee because of their illness (drugs) can lead to legal problems. For those employees being treated for drug addiction, health insurance may pay for a portion or all of the costs.

LIE DETECTORS OR POLYGRAPH TESTING

The Employee Polygraph Protection Act of 1988 generally prohibits employers of commercial organizations from subjecting employees to lie detector tests. These tests are called polygraph tests by professionals and there is an American Polygraph Association which maintains standards of quality and integrity in the use and application of polygraph examinations. Another organization that maintains standards for polygraph testing is the American Association of Police Polygraphists.

There are several exceptions to the act that allow polygraph tests. This means that polygraph tests can be requested for employees of the Federal or Local government, contractors of the Federal government involved in national defense work and employees in certain fields of work that support law enforcement like security guards. These employees can be subject to polygraph examinations under limited conditions.

One other exception to the Employee Polygraph Protection Act is that employees may be subject to a polygraph test when the test supports an ongoing investigation where there was an economic loss to the employer. The conditions under which this exception is allowed are complex and an employer should not attempt polygraph testing without legal review of the Employee

Polygraph Protection Act. There is also an exception that allows polygraph testing when drug manufacture, theft or drug security is involved.

The other common use of polygraph tests is in criminal prosecutions. Persons subject to a criminal investigation can volunteer to take a lie detector test to strengthen their case in an effort to prove their innocence.

Employers should be aware that the Employee Polygraph Protection Act is a federal law. Individual states and communities may have laws that are more restrictive about polygraph testing, but not less restrictive. Therefore, a government or employer may have the right to request a polygraph examination under the federal law, but not under the state law.

The polygraph equipment records a person's physical/emotional response to a series of questions. The equipment records heart rate, respiration and perspiration. Any physical/emotional response to the questions would be recorded by the polygraph instruments. Hence the examiner may discern if the candidate has been untruthful.

Lie detector tests take between 3 and 4 hours when applied by a certified polygrapher. Costs of the tests vary between $250 and $1500 depending upon the locality where the tests are administered and the complexity of the exam. These estimates are for testing by a polygrapher who is a member and certified by one of the associations, either the American Polygraph Association or the American Association of Police Polygraphists. Each association maintains a list of members and will provide, without charge, the names of locally qualified individuals.

STAFF TRAINING

A facility's security is a function of the commitment not only of the security forces but of the entire staff. A cooperative staff, interested in maintaining the facility's security, will help to make the facility more secure by following procedures, keeping doors and cabinets locked, wearing badges and encouraging others to do so. In order to maintain security there are a number of steps a facility manager can take to keep the facility operating smoothly.

When the facility has many security systems like the ones in this book, teaching the staff what security measures are in place and training them to use the systems is as important as the systems themselves. Training starts with procedures. If there are systems in place, then the systems should be accompanied by procedures that indicate how the systems are to be used.

For locks and hardware, a key control policy should be written. The policy should indicate who issues keys, how many keys are issued, how to replace keys and what to do when keys or locks need to be changed. For an electronic lock control system that uses proximity badges, a policy should be in place about how to issue the badges, who has the authority to issue the badges, which doors open with proximity badges, which systems are off limits, etc.

For a CCTV system, there should be a policy about how the system is used and who has control of cameras, what they should record, what to do with the recordings. In the case of a CCTV system, some facilities notify all personnel of the CCTV system policy, in other facilities, management only notifies staff that a CCTV system is present. The CCTV system policies are for those staff that operate and maintain the CCTV system, not for the entire staff.

With these policies in place, the facility can then train the staff as to what the expectations are. Some facilities hand the employee the policy and ask them to read it. In others a log is kept indicating when the policy was given to the employee and the employee is required to sign a certificate that he or she has read and understands the policy. For more complex systems, consulting trainers can be brought in for the express purpose of training employees. These consultant trainers take the policies, prepare written training manuals and course materials. Then using slide shows, photographs, drawings and other tools they teach a class to a group of employees. Classes last from one day up to two weeks. For many security systems, a simple briefing of new employees could suffice. For high security and complex systems, a one to three day class should be sufficient except for a guard force.

In different states, there are rules for the number of hours that are required for training of security guards. Usually this training is about law and law enforcement. Training of guards that carry weapons is more intense and schools that train counter terrorist or police actions are even more intense. These special schools are only for high level security forces and would not be the same as the types of training for the users of systems in this book.

A few facilities may already have a training staff, and if this is the case, then these trainers can perform the same tasks as the consulting trainers. The advantage of the consulting trainers is that they can be brought in to teach the class and then released, perhaps visiting the facility once every six months to a year to update the training of new employees.

Finally, there has been a shift to computer-based training that has been very effective. For large organizations where many employees work in an office setting, a complete training course in security and security policy can be prepared and presented over the internet. The training materials are constructed as a World Wide Web site that includes text, video, audio, slides and even tests. The employees can take the training from their computer workstation. The Untied States Department of Energy has used this type of training effectively. It eliminates travel expenses for the employees. However, computer based training should be limited to about 3 hours, because the effectiveness of longer periods of computer based training isn't as productive.

Depending upon the complexity and sensitivity of the workplace, security training can range from an employee briefing of ten to fifteen minutes for simple tasks for new employees with limited duties, to six weeks for security professionals who manage many elements of a security program including performing risk assessments, hiring and training guards, using background checks and drug screens and the many other elements of loss prevention and control.

ESTABLISHMENT OF A HIGH LEVEL
SECURITY PROGRAM

High level security programs are obviously more complex. These programs include use of physical security systems plus working in or near serious, significant materials (or documents relative to those materials) that could be a threat to human life or the environment. These high level systems include a background check, use of physical security systems, adherence to procedure and complete honesty and integrity on the part of the employees. These types of programs are very rigorous and minor infractions can be grounds for termination from the system.

Generally, after the background check is completed, a certifying official will sit down and chat with the employee. If there are issues from the background check such as previous health problems or minor law infractions, these are discussed with the employee and a determination has to be made whether to allow the employee to continue employment in the high security status.

Some elements are automatically disqualifying such has past illicit drug abuse, criminal activity or alcoholism. In a high level security program, the employee is not allowed to enter into the program, and receive the knowledge of what the assets are, nor the systems to protect the assets. Most organizations do not terminate the employee because of these past records, especially if they were previously disclosed. If the employee has lied and been untruthful, then they can be terminated. Whenever an employee is released for security reasons, the probability of a lawsuit is high. Therefore, it may require legal consultation to determine the safest course of action. The reaction of the disqualified employee might be quite violent and as a result, it may be necessary to utilize the guard force to escort the person off the premises. If there are no guards, it may be prudent to request the assistance of local law enforcement.

Depending upon the size of the facility, there may be one or more certifying officials. These persons should be trained to exercise caution with the background check results since public disclosure of the information may violate the employee's privacy. In general, certifying officials talk in terms of "a candidate" without disclosing the name of the candidate. The decision to disqualify is limited to only those people who need to know.

Once the high level security program is up and running, periodic checking is still required. Many programs require the employee to disclose events that could disqualify them. Some organizations will allow the person to become temporarily disqualified for a few days, weeks or months until the problem is resolved. This has been typical in cases involving drug abuse. The employee is found to be abusing drugs, is warned and given a period of a few weeks or months to get cleaned up. Regular drug testing and counseling is required.

Because of all the checking and training, a high level security program is expensive. Many of the regulations are established by governments who demand from contractors and employees a high standard. Only facilities with significant resources can provide for these types of systems. In general establishment of a new high-level security program for a facility can take up to one year. This amount of time is required for the necessary background checks, evaluation of issues of concern, training of the employees, and establishing a control program they would follow. In most instances, an independent outside entity conducts a surveillance of the program before releasing the facility to operate with the materials of concern.

SECURITY PROFESSIONALS

Chapter 17 provides a list of organizations for security professionals. These organizations provide general policy guidance and centralization of knowledge of past and emergent security issues. A facility that includes a loss prevention or corporate security organization may place this organization within the human relations organization or it may be a part of the business organization. The manager of this element is sometimes referred to as the Chief of Corporate Security or the Chief Loss Prevention Officer. Often these persons have had a background in law enforcement and are former police officers, detectives, or criminal investigators. In some organizations, they can have a background in human relations or facilities management.

Several national and international organizations offer an accreditation based upon a candidates skills, knowledge and experience. The organizations differ slightly in requirements.

CERTIFICATIONS

Several organizations certify the expertise of people in the security profession. However, as of this date, it does not appear that there is a legal requirement to license a security professional in the same way that there is for the legal profession or for law enforcement. A certification is a voluntary credential issued by an organization of like minded professionals. In many cases, members of a state board are also members of the organization that issues certifications. Certifications are designed to assure that a person has fundamental knowledge of security and loss prevention.

AMERICAN SOCIETY FOR INDUSTRIAL SECURITY (ASIS)

One organization that provides certifications of security professionals is the American Society for Industrial Security (ASIS). ASIS has three tiers or levels in their certification program: The Certified Protection Professional (CPP), the Professional Certified Investigator (PCI) and the Physical Security Professional (PSP).

Certified Protection Professional or CPP

The certified protection professional is the most difficult level of certification to be obtained from ASIS. In order to qualify, a candidate must have at least 9 years of security experience, of which 3 years they have been in charge of a security function. For candidates with a bachelor's degree from an accredited university, the practical experience time is 7 years. Three years in charge of a security function remains a requirement. No criminal convictions are allowed for a candidate for the Certified Protection Professional. After application, it is necessary pay a fee for the background check and for an examination. The exam consists of a multiple choice test of approximately 200 questions broken down into broad categories relative to security management. These categories include: management and supervision, risk and vulnerability assessments, policy development, personnel management and loss prevention. The next category, investigations, includes types and methods of conducting investigations and the resources necessary for investigations. A third category, personnel security, includes questions about personnel matters as they pertain to security, including selection and training of personnel, background checks and disciplinary actions. A fourth category, physical security, includes questions about employee and visitor control, alarms, barriers, electronic devices, lighting, perimeter security, guards and guard patrols and weapons. The fifth category, sensitive information, includes questions about identifying sensitive information and controlling it. The final category, emergency management, includes questions of types of emergencies, planning for emergency response and implementing emergency response plans.

Upon completion of the application, background check and payment of fees and after successfully passing the exam, a certifying board meets to discuss the candidates' credentials. By successfully completing the process, the candidate meets basic requirements for security management.

Physical Security Professional

Like the Certified Protection Professional, ASIS also provides a certificate for a physical security professional (PSP). It is not necessary to become a Physical Security Professional in order to advance to a Certified Protection Professional, but the requirements for the Physical Security Professional are not as difficult to achieve. Less experience is required for the Physical Security Professional with 5 years of experience in physical security and a minimum of a high school diploma. No criminal convictions are allowed for this certification. After application and paying the appropriate fees, a candidate will be given a multiple choice exam that covers three basic physical security areas. They are: 1) performing a physical security assessment, 2) proper selection of integrated physical security measures, and 3) implementing those security measures. Part 1, the assessment, includes those elements essential to risk assessment and performing a risk assessment. Part 2, selection, includes questions about understanding what types of equipment and systems can be installed or implemented to protect the assets and pricing methods for selection of alternates. Part 3, implementation includes a basic understanding of the various methods to provide the systems and equipment identified in Part 2. This would include bid and bid evaluation, project management, training, testing and acceptance, maintenance and loss prevention.

Professional Certified Investigator (PCI)

ASIS also provides a third certificate, the professional certified investigator. This certificate is more closely related to law enforcement in that it includes experience case management, evidence collection and presentation.

THE AMERICAN COLLEGE OF FORENSIC EXAMINERS INSTITUTE (ACFEI)

In addition to the certifications available from ASIS, the American College of Forensic Examiners Institute (ACFEI) has initiated a program for certification in homeland security (CHS). Certification in Homeland Security, like the ASIS programs, has been created for the certification, training and continuing education of Homeland Security professionals. Because this program is relatively new, candidates who have extensive previous experience in law enforcement, fire fighting or emergency medical services will be given credit for that experience in the certification program. In addition to previous experience, membership in the American College of Forensic Examiners Institute and no previous criminal convictions allows a candidate to be eligible for certification. Membership in the Institute costs $135 per year and the candidate must agree to supporting continuing education in homeland security issues. In addition to the ACFEI annual dues and the requirement for annual training, the certification program costs $350.

CRIME PREVENTION THROUGH ENVIRONMENTAL DESIGN (CPTED)

In addition to the American Society for Industrial Security and the Certification in Homeland Security there is a third group that sponsors Crime Prevention through Environmental Design (CPTED). The Association is called the International CPTED Association or ICA. The ICA offers a certification as a "CPTED Practitioner, ICA Certified" (ICCP-advanced or ICCP-basic). In order to become certified a candidate must demonstrate skill and knowledge in core competencies of crime prevention through fields like architectural design, urban planning or social planning that are relative to this association's certification. The candidate must apply for certification with previous experience in environmental design as it relates to crime prevention and after a review by the agency, International Crime Prevention through Environmental Design Association (ICA), the candidate is subjected to an exam that covers core principles. Because the ICA is broader in scope than ASIS, its exam structure is more flexible. For persons with different skills, the exam may be oral or written. The written exam is essentially a 10 page document that responds to written questions from the certifying board. Candidates are given a week to prepare their answers in a take home format. The cost for this program is $200.

SYSTEM TESTING

Any installed system, whether it is an access control system with electronic door locks or a CCTV system, requires periodic testing and maintenance to assure it performs in proper order. A new system should be thoroughly tested after installation. Older systems should be periodically tested to maintain their working order.

For new systems, performance testing can be performed by the one of three groups—the vendor, the installing contractor or the facility. Each method has advantages and trade-offs. Allowing a vendor to conduct the performance testing means that the system will be tested and obvious flaws quickly repaired. However, vendor tests can end up incomplete when flaws are identified because some of the materials or components are not compatible with the vendor's system.

For the installing contractor to perform the tests, a similar situation can develop. The interfaces between components can fail and instead of fixing these interfaces the installing contractor claims that this was not a part of his original scope. He may claim the work is extra for which he expects additional payment.

Finally, the facility can perform the system tests, which in the long run is the most effective, but this can take valuable staff time that was not in the original facility budget. By having the test conducted by the facility staff, the staff will be most familiar with the system upon completion. If they have had to troubleshoot some systems, then they will be capable of trouble shooting problems during system operation.

Some vendors elect to perform a factory acceptance test or FAT. This type of test is run in a factory or large facility where all of the systems are set up and tested, and their relationships checked with the other components before shipping the system to the field for field installation. After the equipment is installed, a site acceptance test (SAT) can be conducted. Finally, an operational acceptance test (OAT) can be conducted after all systems are completed and all operators are fully trained.

Testing means each device in the system is operated and verified to perform as designed. Elaborate methods have been developed to eliminate confusion or failure during the test, so a Facility Manager needs to make sure that the testing is, in fact, providing the necessary confirmation that the systems are working correctly.

The first step in field testing is to verify that the signals have been run correctly to the devices. This is sometimes called "shooting-the-loops." Shooting-the-loops means that the signal is sent out to the device and the device responds. This does not mean that the device actually does what it is supposed to do yet, it just means that all the wires are in the right places. Cables to the devices might be run through several boxes that act as pull stations along the lines. If one of the wires is incorrectly landed, the device won't respond properly. Once the wiring is validated by shooting-the-loops, then a function test of that device is performed.

Does a camera pan, tilt and zoom as intended? Can it be switched on and off? For an access control door that works with proxy badges, does the database actually recognize the badge? What if a person brings in a proxy badge from another facility? Does it work? Is it supposed to?

All these types of questions need to be addressed when setting up the system's functional test plan. Once the staff performing the tests agrees with the plan, then the plan is taken into the field and tried. What are the criteria by which the tests pass or fail? The contract for the system needs to state that the test will be conducted and if the test fails, the system will be corrected and the test repeated until the system passes the tests.

Contractor finds equipment testing loophole.

On one facility the installing contractor repeated the failing test but the contract failed to stipulate the test would be repeated until it passed. When the test failed the second time, the contractor stopped testing and insisted he met his contract. A second contractor had to be hired to finish the job of the first contractor. Payment for the work was settled after a long court battle.

MAINTENANCE

On occasion, a device in the system will fail. Cameras go out, readers stop reading, locks will not hold, etc. When this happens, the device has to be repaired or replaced. The replacement device for a new system is usually the same make and model as the old one, but as time goes on, newer devices make the old ones obsolete. When a new device is installed in a system, it too should be tested for compatibility, in the same way the original was tested.

When the tests are completed, the system is ready to support the security mission again.

One of the most important parts of a security system is the people who operate and work within the system. It is extremely difficult to protect assets if people do not care whether the assets are protected. By "selling" security, helping employees to understand the need for security systems, badging them for identification, and checking backgrounds, a facility can greatly improve its asset protection programs.

When staff is needed, a security professional with professional credentials can be hired to help improve a facility's security.

Finally, systems should be constantly tested and diligently maintained to assure that the systems work properly when needed. A facility needs a security system the most when an incident occurs. The next chapter, Emergency Response, is the final step in assuring a good security system works as planned.

RESOURCES

MANAGEMENT and TRAINING
The Dale Carnegie Training System offers courses and seminars in
the core areas of Leadership Development and Systems
www.dalecarnegie.com
DRUG TESTING
Drug Testing Policies on the Web can be located at www.ohsinc.com
Drug Use: The White House Office of National Drug Control Policy
Drug Data Summary, March 2003.
www.whitehousedrugpolicy.gov/publications/fascsht/drugdata/
index.html
BACKGROUND CHECKS
Local and National Criminal Background Checks are available from:
Choice Point http://www.choicepoint.net/industry/retail/
pre emp 4 1.html
Search Systems http://www.searchsystems.net/
Washington Research: http://www.backgroundcheckgateway.com/
John E. Reid and Associates
Offers Seminars, Textbooks and Training on Interviewing,
Interrogating and Lie Detection Techniques www.reid.com
Polygraph Testing (Lie Detectors)
The American Polygraph Association www.polygraph.org.
The American Association of Police Polygraphists
www.policepolygraph.org/
United State's Code, Title 29 Chapter 22 Employee Polygraph
Protection Act
Badging Equipment Vendors:
Data Card Systems www.datacard.com
CERTIFICATIONS
American Society for Industrial Security or ASIS Certifications guide-
lines is located on the web at www.asisonline.org and following

the links to Certifications.

American Board for Certification in Homeland Security
www.acfei.com

International Crime Prevention though Environmental Design Association (ICA) www.cpted.net

Chapter 16

Emergency Response

Emergency response is a facility's ability to respond to an emergency situation. If a facility determines from the risk assessment that a security system is necessary, then in addition to whatever security measures exist to protect the assets, the facility should have the ability to deal with an emergency concerning those assets. This chapter is provided to assist in the development of an emergency response plan and provide guidance into the steps that are normally taken in dealing with an emergency.

The function of security is to prevent an incident, but in order to provide an effective deterrent—a fundamental assumption must be that security will successfully deal with that event. Otherwise, the security program is ineffective. How a facility prepares for the emergency and how people perform in an emergency is the subject of this chapter. Depending upon the type of facility, the staff should either be trained to respond to an emergency or they should have confidence that others have been trained to respond in an emergency and their efforts will be guided and directed by that group.

It is probable that an emergency response plan will reduce liability premiums. For some facilities dealing with special hazardous materials, an emergency response plan is required by community or federal regulation.

EMERGENCY RESPONSE PLAN

The first step in preparing for emergency response is to have an emergency response plan. The plan defines what an emergency is, what resources are available for dealing with it, who is in command and how the resources will be directed. The plan usually includes testing of the plan (conducting emergency response drills) where the facility's resources are tested in mock incidents or emergencies to confirm readiness.

Hazard Assessment

Preparing a hazard assessment is relatively straight forward if the assets and types of facility have been developed as a part of the risk assessment. A hazard assessment includes threats from hazards other than security risks. While most security events are man-made, the hazards for the emergency response plan include natural hazards that could threaten a facility. Natural hazards include weather related incidents like a tornado, hurricane and flood. Other natural events that are not weather related include earthquakes or wildfires. These natural hazards are a function of the location of the facility. For a facility that is located within the boundary of a local government, some emergency response planning for natural disasters may have already been done for the community. Facility managers can check with their local emergency response organization which can be located through city, county or state government. Each locality is different and some function responsibility for these hazard assessments has been changing as the U.S. Federal Government centralizes emergency preparedness under the Department of Homeland Security.

Command Structure

The next element in the emergency response plan is establishment of a command structure. The command structure indicates who will take charge and what decisions they will make in an emergency. Most facilities have a call down list that employees are expected to use if an event unfolds. The call down list starts with the name of the first person to be notified, then the next, then the next and so on. For elaborate facilities there is a form with the names and phone numbers of the persons on the call down list. Some facilities designate a team called the crisis management team or in others it is called the incident command system. However, the main point is to establish a command structure to manage the resources during an emergency.

Designation of leadership in the command structure of an emergency response plan should not be taken lightly. These persons may have to make life or death decisions.

In an unfolding event, the command structure will not be intact when an event starts. The authority to establish an emergency event and to make decisions should be defined. The command should also know who to notify, and be able to coordinate resources during a response. For example, in a few highly trained facilities, the command structure talks with the fire department via two-way radio while the fire department trucks are en route.

Communications

The next element of the emergency response plan is to define how communications will be established and maintained. There are three elements to the communications part of the emergency response plan.

First, communications should indicate how an emergency event is established. The command structure will not be established until after the event commences, so how does a person notify the command structure of an event? This is the reason for the call down list in the command structure. Who is to be notified and when? What method will be used for notification? For most facilities a telephone is used. Some facilities have a loud speaker paging system; others use voice notification and a runner. Finally, most facilities are equipped with alarm systems, such as the fire alarm system. A person can trigger a fire alarm if the building is on fire, for example.

The second type of communication is notification of the request for assistance. For a fire alarm, this is relatively simple. Signaling a fire alarm will bring the fire department. They have the equipment and training to deal with many events. However, pulling the fire alarm in response to a different type of event could have adverse consequences. For example, a bomb threat may include explicit instruction not to initiate a fire alarm. The plan needs to consider alternate ways to communicate depending upon the hazard.

Finally, the communication plan should include a method of notifying the facility's employees what they should do in response to the emergency. In some cases they would evacuate, in others, they may need to gather in a sheltered area of the building. The communications plan needs to include the method of notifying the staff of the actions they should take.

Accountability

The plan should indicate how to account for the people in the facility. If there is an evacuation, there needs to be a method of verifying everyone is safely evacuated. Some facilities have persons designated as sweepers. Their role is to go through the facility and verify that everyone has been safely evacuated. Another method is to perform a head count or identification of the persons evacuated to determine if all have been evacuated. This can be very important as rescue personnel may have to enter a burning building to look for someone. Chapter 9: Computer Access Control explains the mustering option available from some software vendors to aid in establishing accountability.

Training

The emergency response plan should include a method of training persons in how to respond to an emergency. There are a number of methods of training from written guidelines to emergency response drills. Drills are important and the organization capable of responding to drills has a better chance of adequately responding to an actual event.

Procedure Training

There are a number of training methods that should be included in the emergency response plan. First, employees should be given the procedure and should be encouraged to read it. Second, for a large facility, there may be a corporate training department. If this is the case, employees can be trained in a classroom. In addition, consultant trainers can be hired to train employees at the facility or the staff can travel to where the training is taking place. This type of training can quickly become expensive, especially if travel is involved. Independent corporate trainers can come to the facility for less money if the class size is about 5 or 6 or greater. Another method to train staff is computer based training.

What happens with computer-based training is that the course materials are created as a combination of video, audio and text in a format similar to a slide show. The employees then participate in the training from their computer work station. One nice thing about this type of training is that the audio portion of the lessons can be broadcast through the computer's speakers. This way an employee can hear the various alarm sounds without having to set them off in a drill scenario. One problem with this type of training is that it takes work to make it facility specific. But once done, it can be used over and over to train and retrain employees. For a large organization, this may be acceptable but for a small organization, computer based interactive training may be so expensive it becomes infeasible.

Tabletop Exercise

The next step in the training part of the emergency response plan is called a tabletop exercise. This type of exercise consists of the command structure working together at a table, talking through what they would do in a mock emergency response exercise. Usually, a tabletop exercise is recommended before trying to go into a full blown operational exercise as the processes and resources are only theoretically committed.

Drills

Finally, the plan should include conducting the ultimate training exercise: the emergency response drill. The emergency response drill includes operation of the facility alarms and staff responses to those alarms. The emergency response drill should include the notification of the command structure, use of the communications equipment and use of the command structure to disperse the various resources. The drill should also demonstrate the establishment of accountability by whatever method the facility has chosen.

Revisions

The emergency response plan must be revised when conditions change, when new hazards are recognized or when staff turnover has changed the makeup of the command and control structure.

Once the plan is established the next step is to train the staff and members in their actions in an emergency.

Media

One of the most difficult elements to deal with in a real event is the response of the media. When the media is notifying the public, which they have every right to do, the public response can easily cause the communications equipment to become over-whelmed by phone calls jamming up the telephone lines. For some facilities a media room is provided. The room has multiple telephones and internet connections to allow media to communicate their reports to their networks. It is up to the facility to determine whether to include dealing with the media as a part of the emergency response plan. Some facilities invite the media to participate in emergency response drills to test the equipment. In a real event, however, whatever was established for the media beforehand can quickly go awry.

For facilities that allow the media to participate, they usually have a protocol or public relations officer deal with the media during the event.

CRISIS MANAGEMENT SOFTWARE AVAILABLE

In an effort to help local government compare the features of numerous crisis management software products the National Institute of Justice, National Institute of Justice/Office of Science and Technology prepared a Crisis Information Management Software (CIMS) Feature Comparison Report. The report tested 10 vendor products of Crisis Information Management Programs. The Report Evaluation included the preparation of an evaluation matrix that is downloadable from the web and runs in Excel® format. The user inputs elements specific to the needs of their facility and the matrix evaluates and compares the products. The report is located on the World Wide Web at:

http://www.ncjrs.org/pdffiles1/nij/197065.pdf
and the matrix for evaluation is located at:
http://www.ojp.usdoj.gov/nij/temp/publications/197065-Matrix.zip.

Use of the matrix requires response to over 200 questions about services, databases, functions and preferences. The output is provided in the form of a table indicating the weighted rankings of the various programs. Figure 16-1 is a partial printout of results.

EMERGENCY RESPONSE TEAM

A facility may decide to establish and emergency response team. Depending upon the nature of the emergency, this may be a team trained to respond to a hazardous materials incident (HAZMAT), a team to respond to an incident using special weapons and tactics (SWAT), or a bomb squad. Not all of these resources are needed for each facility, but these are the types of teams that a facility may use. If these facilities are available, then drills should be scheduled to exercise these teams to maintain proficiency. The emergency response team should be given the equipment they would need in responding to an incident and then they should be trained to use it. Special teams are costly because their training requirements are higher than the rest of the staff. And special teams often require higher wages because of their elevated level of expertise and training. In addition, regulations usually require these teams to be larger than expected to provide backup for each other. For example a HAZMAT team needs a backup team to be ready for rescue if problems develop with the initial team.

In a few scenarios, more than one type of team may be necessary to adequately respond, further complicating the communications resources and overall response.

DRILLS

After establishing a plan, training, practicing with the command structure, using the communications devices and preparing the emergency response teams (including equipment) the facility is ready to conduct a drill. Drills are very important because the lessons learned in the drill are what the responders will use to respond to an actual event. There is no guarantee persons in a real event will actually perform as they have been trained, but without practice, unfolding emergency events can have disastrous consequences. Many drills fail in the initial stages because the first response is incorrect and all steps after that are made more complex as a result of the initial errors.

Drills should only be conducted as necessary. For many facilities, there is one annual drill, usually an evacuation. In some facilities, like schools, there is a drill every semester. For new facilities, more drills are required than after the facility has been operating for a short period. Initially drills should start with less complex scenarios and progress to more difficult scenarios. Some facilities miss this point and try to save money by conducting the complex drill first. Some drills unravel so badly that nothing is learned.

The Drill Controller

To properly set up and conduct a drill, the organization should designate a person to control the drill scenarios. The designated person is usually called the *controller*. There should be only one controller as multiple controllers tend to make the drill unwieldy. Assumptions have to be made during the drill and it is the controller who should say what may be assumed and what is not assumed.

Drill Evaluators

In addition to the controller, the drill may need evaluators. Evaluators are persons who observe what happens during the drill and evaluate the performance of multiple elements. The controller can be at the scene of the event and evaluators at the command center or the evaluators can be at the scene of the event and the controller at the command center. The persons responding to the drill are "players." They are the ones who are being evaluated. Finally there may be a need for persons who must operate the facility while the drill is going on—these persons are "exempt" from the drill. Part of the duty of the controller is to determine who is a "player" and who is "exempt."

For complex drills, different staffs may wear armbands indicating their status. Finally, for some drills there are "aggressors." The aggressors are persons who provide the impetus for the drill. In security, the aggressors play mock terrorists or intruders. Drills with aggressors must be carefully controlled as young people performing drills can get slightly carried away and someone can actually become injured.

In all drills, it is important to remember that the drill is an exercise. Should real events emerge, the drill should be canceled immediately.

Objectives

Each drill should be designed with an objective in mind. The drill should be designed to test and demonstrate one or more elements of the organization's ability to respond to an event. The scope of the drill should be planned by the controller, usually in conjunction with management. The controller and the evaluators should meet before the drill and discuss the scope of the drill and what elements are going to be tested during the drill. The evaluators should know what the objectives are, and they should score the drill against the objectives. While there is always the possibility that an event could occur that would require more resources than the facility has, testing should be set to stretch the resources, not overwhelm them. As an example, a scenario that has 40 or 50 casualties in a facility that is capable of treating 4 or 5 casualties would obviously fail. Creation of a scenario that fails is not a wise use of the resources.

Drills should be designed for minimum of intervention by the controller. The best drills can be accomplished with something that starts out small and then the scenario is allowed to unfold as it would by allowing the players to react as they normally would. The more a controller has to intervene in the interpretation of the scenario by the players, the more difficult the evaluation of the drill will be. Sometimes players can't tell what part of the drill is, and what isn't.

One element tested extensively during a drill is communications. How do the radios work? Who responds on the telephone? Did they receive the correct information? Did they react appropriately? Was the information presented accurately?

Another element to be evaluated is the decision making of the command team. Were their decisions appropriate? Were their communications clear? How did they handle changing conditions? Finally, management needs to know enough about the decision making process that they do not attempt to second guess the command structure in a real event. The command structure has extreme difficulty dealing with an unfolding situation. For someone who is not in command to try to assume command is a mistake. But, in the real world, this is one of the problems the command structure will face.

Depending upon the facility, coordination with multiple resources may or may not be required. For example, if the fire alarm is used, the fire department should be notified that a drill is going to take place. If law enforcement or medical support is used, these resources should be notified as well. Depending upon the resources, these groups may become players in the drill in an effort to show community or area ability to respond.

During an emergency response exercise a mock casualty was being evacuated from the building when the mock casualty had a mock heart attack that was included as a part of the drill. The Emergency Medical Team responding to the casualty took out a defibrillator and indicated that they would apply it to the victim. The lead EMT called out "Clear!" and proceeded to pretend to shock the victim. At this point the controller of the incident stepped up to the stretcher and said, "You were not clear," pointing to another EMT who was attending to the victim. The controller then said, "You are now a casualty also, lie down and wait for aid." The EMT wasn't very happy, and this doubled the problem for the team being drilled, but in a real situation, if the EMT had not cleared the stretcher, he, too, would have become a casualty.

Drills are Serious Business

The support of the staff to a drill is directly related to how serious management takes them. And finally, in a real event, everyone who has ever been in a drill agrees they were glad they got to drill before they were exposed to the real thing.

A real event may be something never before seen. Making up a response on the fly is very difficult. When an unforeseen event occurs, everyone must do their best. If a manager has to make life or death decisions, he will be glad he or she drilled when they got the chance.

Evaluation

After the drill is concluded there should be an after action review. The review should state the objectives of the drill and how successful the facility was in responding. The evaluation should be positive, even if the facility failed to meet the objectives. If something goes poorly in a drill, it often happens at the very early

In drills involving weapons, the aggressor weapons are not actually weapons at all. Sometimes they are mock weapons cut out of wood, or they are painted red, or both. The point being that no mock aggressors should have functioning weapons. The players should not have any ammunition and the boundary between the exempt guard force and the drill force (players AND aggressors) should be defined and carefully controlled.

In addition, the stand down from these types of drills should be carefully controlled. Weapons should be double checked. And when the drill is over and the force resumes normal guard duty, it should be clear to everyone that the drill is over.

stages of the drill before anyone has figured out a drill is in progress. This is the same thing that happens in a real event. Valuable lessons can be learned from drills. If a facility works multiple shifts, the results of the drill should be presented to all shifts giving them the opportunity to learn as well.

More about Drills

Drills can be announced or unannounced. Usually a new facility will announce drills until the facility has been successful and then start unannounced drills. Often, drills take place on day shift, because this is when most of the people are present. Some dedicated facilities conduct drills on evening shift, holding over key people from day shift to act as evaluators and support resources. Very few drills are conducted at night. In general, drills are not conducted first thing in the morning, facilities choose instead to wait until 10:00 a.m. or even until after lunch. Drills rarely continue into night shift—the controller ending the drill at quitting time allowing everyone to go home.

Real events, however, start at night or on holidays. Chernobyl started at midnight, and the situation was hopelessly out of control before anyone in the command structure was notified the next morning. Natural disasters occur at odd hours. Teton Dam, the large dam that failed in Idaho, failed on a Saturday morning. Pearl Harbor was attacked at 7:00 a.m. on a Sunday

Example Form for
Emergency Response Drill Evaluation Report

Emergency Response Drill Evaluation Report   Example

Date of Drill _____   July 25, 2005

Time from: _____ To: _____   1300 to 1700 hours Local

Name of Facility _____   Washington High School

Drill Scenario _____   Ex. Natural Gas Leak/Basement

                _____

                _____

Initial Notice Discovered by _____ Name of Person

      Initial Response: _____   What did this person do? Call.

Notification Received by: _____ Name of Person

Action Taken _____ Verified Location and Notified Management

Command Notified Time_____

      Person Notified _____

Emergency Response Team Notified (In this case the fire department.)

Evacuation Required? Yes ____ No _____ (Depends upon the response.)

Accountability Attained? Yes ____ No ____ Time Attained _____

Time event was controlled. _____

Controller Interventions Required? Yes ____ No _____ Example: Yes

   Intervention _____ Example: (Fire Department was not

      _____ notified.)

      _____

Evaluator Comments:

      _____   Fire department telephone number

      _____ was not available.

      _____

After Action Review _____ Meeting held with team.

      _____

      _____

This drill passed? Yes _____ No _____   Example: No.

   Reasons _____   Accountability could not

      _____   be attained.

      _____

Lessons Learned from this drill

      _____   Example: Post fire phone

      _____   department numbers

      _____   prominently.

morning. My personal favorite time for a drill is at about 7:30 p.m. on a Sunday evening.

For some scenarios, certain elements of a drill have to be exempt. For example, in security drills, all the live ammunition has to be locked up. Live fire drills should never be conducted. Some types of operating equipment have to continue to be operated during the drill even though the drill scenario assumes that the equipment has failed. This is where a controller can help with designing a good drill. Those elements that continue as planned should be considered exempt.

The scenario should recognize that a stray element may wander into the drill area and the controller should determine ahead of time what to do with the drill if this happens. Sometimes, these persons have to become players and sometimes they can be exempt. It is the controller's call, but he should consider it before the drill starts and tell the evaluators.

Drill Safety

Finally, safety should be very important during a drill. People have been accidentally shot during a drill, people have passed out during a drill, and people have become so stressed during the drill that they have a heart attack and have died as a result of the drill. This, of course, should never happen so whenever a facility conducts a drill, safety should be emphasized.

On night shift at a military base in Utah in the early 1990's a drill was scheduled. This drill required personnel to put on their gas masks and proceed to the assembly point. One of the workers, stressed by the event, could not breathe enough air through the valves in the gas mask. He fainted and then had a heart attack. The team had difficulty converting what was supposed to be a drill into a real event. It took precious moments to get the command structure to switch from drill thinking to real thinking. The scenario did not include any casualties so a sudden heart attack victim was unexpected. The command team initially did not know that the heart attack was not a part of the drill. The legal complications of this event went on for years. When a drill is conducted the team should recognize that the events from the drill may cause the event to switch to a real event.

At one facility the safety oversight representative was asked to verify the limiting conditions of operations. This was a document the facility maintained to assure that resources, (people and systems) were available to deal with an emergency if one occurred. The limiting conditions of operations included a requirement for a trained and certified on-scene incident commander be present. This person, normally the shift manager, took the day off and was not present. The deviation to the limiting conditions was to allow operations to continue with an alternate person serving as the on-scene commander if there was an emergency. The proposal was that the subordinate, the chief of operations; and the shift manager's supervisor, the site manager, were both trained and certified. Either of the men would fill in as the on-scene commander if something happened. The safety representative saw the two men sitting in the control room, so he asked them, "Ok, which one of you is the on-scene commander if an event occurs?" They each pointed at the other one and said, "He is." The safety representative looked at the two men for a moment...

Then the operations manager said, "Ok, I'll be the on-scene commander and if I'm not here, then the site manager will do it."

If an emergency had occurred, immediate confusion would have ensued because the on-scene commander role was undefined.

READINESS

With the drills complete and lessons learned, the staff is trained and ready for an emergency. The facility is as secure as it can be, it has a trained staff. Security systems have been tested and are known to perform. The risk assessment is current and up to date. The assets are protected.

RESOURCES

Gustin, Joseph F. *Disaster and Recovery Planning: A Guide to Facility Management 3rd Edition*, Lilburn, GA; The Fairmont Press 2004 www.fairmontpress.com

CIMS Feature Comparison Matrix

All Feature Results
These are the results based on your entered criteria.
Click on Vendor Name to get detailed scoring

	Blue292 2.0	Crisis 5.2	EM2000 3.1	E-Team 1.6	Incident Master 1.5	OpeCenter 2.3	RAMSAFE 2.0	RESPONSE 8.0	SoftRisk SQL 5.1	WebEOC Profess'al 5.3
System Environment										
ASP	177.00	N/A	N/A	217.50	N/A	202.00	N/A	N/A	N/A	188.50
LAN	N/A	163.50	157.00	209.50	157.00	206.50	190.50	178.00	158.00	180.50
Hybrid	N/A	N/A	N/A	424.50	N/A	N/A	N/A	N/A	N/A	374.00
Functional										
Reviewed Score	ASP	LAN	LAN	LAN	LAN	LAN	LAN	LAN	LAN	LAN
Product Support										

Figure 16-1. Partial print of crisis information software comparison report. Source: National Institute of Justice, National Institute of Justice/Office of Science and Technology prepared a *Crisis Information Management Software (CIMS) Feature Comparison Report*.

National Institute of Justice: Crisis Management Information Software Test Evaluation.

http://www.nlectc.org/virlib/InfoDetail.asp?intInfoID=605

US Federal Government Guidelines for Emergency Response and Preparedness are located on the World Wide Web at: http://www.osha.gov/SLTC/smallbusiness/sec10.html

Chapter 17

Security Associations: Who Can You Turn To?

The American Association of Automatic Door Manufacturers
    AAADM
1300 Summer Avenue
Cleveland, OH 44115
215-441-7233
www.aaadm.com

The American Association of Police Polygraphists
18160 Cottonwood Rd # 253
Sunriver, Oregon 97707
888-743-5479
Fax: 541-593-1021
www.policepolygraph.org

American Board for Certification in Homeland Security
American College of Forensic Examiners International
2750 East Sunshine
Springfield, MO 65804
800-423-9737
Fax: (417) 881-4702
www.acfei.com

The Associated Locksmiths of America, Inc.
3003 Live Oak Street
Dallas, TX 75204
214-827-1701
www.aloa.org

The American Polygraph Association
APA National Office
PO Box 8037
Chattanooga, TN 37414-0037
1-800-APA-8037 or (423) 892-3992
fax (423) 894-5435
www.polygraph.org

ASIS-American Society for Industrial Security
1625 Prince Street
Alexandria, VA 22314
703-522-5800
www.asisonline.org

Dale Carnegie Business Group, Headquarters
2121 Argentia Road, Suite 103
Mississauga, Ontario L5N 2X4
Tel: (905) 826-7300/1-800-361-2032
Fax: (905) 826-5565
www.dalecarnegie.com

Department of Homeland Security
Washington, DC 20208
www.dhs.gov

The Door Hardware Institute
14150 Newbrook Drive
Suite 200
Chantilly, VA 20151
www.dhi.org

The Espionage Research Institute
240-273-8823
www.espionbusiness.com

Federal Bureau of Investigation
www.fbi.gov

Federal Emergency Management Agency (FEMA)
500 C Street, SW
Washington, DC 20472
202-566-1600
www.fema.gov

International Association of Professional Security Consultants
  (IAPSC)
1444 I Street
Suite 700
Washington, DC 2005-2210
www.iapsc.org

International Crime Prevention Through Environmental Design
  Association
Box 12 Site 17 RR2
Strathmore, AB T1P 1K5
Canada
www.cpted.net

International Security Management Association
P.O. Box 623
Buffalo, IA 52728
800-368-1894
www.ismanet.com

Joint Commission on Accreditation of Healthcare Organizations
One Renaissance Blvd.
Oakbrook Terrace, IL 60181
630- 792-5000
www.jcaho.org

National Alliance for Safe Schools
Ice Mountain
P.O. Box 290
Slanesville, WV 25444-0290
304-496-8100
www.safeschools.org

National Burglar and Fire Alarm Association (NBFAA)
8300 Coleville Road
Suite 750
Silver Spring, MD 20910
301-585-1855
www.alarm.org

National Crime Prevention Council
1700 K Street, NW
Second Floor
Washington, DC 20006-3817
202-466-6272
www.ncpc.org

National Fire Protection Association
1 Batterymarch Park
P.O. Box 9101
Quincy, MA 02269-9101
617-0770-3000
www.nfpa.org

Pacific Northwest National Laboratory
P.O. Box 999
Richland, WA 99352
888-375-PNNL (7665)
www.pnl.gov

Security Industry Association (SIA)
635 Slaters Lane
Suite 110
Alexandria, VA 22314
703-683-2075
www.siaonline.org

Service Employees International Union SEIU
1313 L Street, NW
Washington, DC 20005
202-898-3200
www.seiu.org

Bibliography

Gladwell, Malcolm. *The Tipping Point: How Little Things Can Make A Big Difference*, Little, Brown and Company, 2000

Kovacich, Gerald & Kalibozek, Edward P. *The Manager's Handbook for Corporate Security* Burlington, MA: Butterworth-Heinemann 2003

The National Electrical Code® NFPA 70 published by the National Fire Protection Association, Quincy, MA 2004.

Ohio, Denise *Five Essential Steps in Digital Video: A DV Moviemaker's Tricks of the Trade* Que Corporation, 2002.

Owen, David D. & RS Means Engineering Staff, *Building Security: Strategies and Costs* Massachusetts: Reed Construction Data 2003

Phillips, Bill *Locksmithing* New York: McGraw-Hill a Division of McGraw-Hill Companies, 2000

Physical Security US Army Field Manual FM 3-19.30 HQ TRADOC. Commandant, US Army Military Police School (USAMPS), Fort Leonard Wood, Missouri 65473-8926.

Reference Manual to Mitigate Potential Terrorist Attacks Against Buildings FEMA 426 Federal Emergency Management Agency, Department of Homeland Security; December, 2003.

Reid, Robert N. *Roofing and Cladding Systems Handbook: A guide for facility mangers* Lilburn, GA: Fairmont Press 2000

Security and Safety Measures (Brochure) Vance International, Inc. Oakton, VA

Shannon, M.L. *The Bug Book: Everything You Ever Wanted to Know About Electronic Eavesdropping...But Were Afraid to Ask* Boulder, Colorado, Paladin Press 2000.

Traister, John E. & Terry Kennedy *Low Voltage Wiring, 3rd Edition*, McGraw-Hill, NY, 2003

Whidden, Glenn H. *The Attack on Axnan Headquarters: An Espionage Operation, 2nd Edition*, 1st Books www.1stbooks.com.

Glossary

Access control—Any combination of barriers, gates, electronic security equipment, and/or guards that can deny entry to unauthorized personnel or vehicles.

Access control point (ACP)—A station at an entrance to a building or a portion of a building where identification is checked and people and hand-carried items are searched.

Access controls—Procedures and controls that limit, or detect access to, minimum essential infrastructure resource elements (e.g., people, technology, applications, data, areas/or facilities), thereby protecting these resources against loss of integrity, confidentiality, accountability, and/or availability.

Access Control System (ACS)—Also referred to as an Electronic Entry Control System; an electronic system that controls entry and egress from a building or area.

Accountability—The explicit assignment of responsibilities for oversight of areas of control to executives, managers, staff, owners, providers, and users of minimum essential infrastructure resource elements.

Acoustic eavesdropping—The use of listening devices to monitor voice communications or other audibly transmitted information with the objective to compromise information.

Aggressor—Any person seeking to compromise a function or structure.

Airborne contamination—Chemical or biological agents introduced into and fouling the source of supply or breathing or conditioning air.

Airlock—A building entry configuration with which airflow from the outside can be prevented from entering a toxic-free area. An airlock uses two doors, only one of which can be opened at a time, and a blower system to maintain positive air pressures and purge contaminated air from the airlock before the second door is opened.

Alarm assessment—Verification and evaluation of an alarm alert through the use of closed circuit television or human observation. Systems used for alarm assessment are designed to respond rapidly, automatically, and predictably to the receipt of alarms at the security center.

Alarm priority—A hierarchy of alarms by order of importance. This is often used in larger systems to give priority to alarms with greater importance.

Annunciation—A visual, audible, or other indication by a security system of a condition.

Area lighting—Lighting that illuminates a large exterior area.

Assessment—The evaluation and interpretation of measurements and other information to provide a basis for decision-making.

Asset—A resource of value requiring protection. An asset can be tangible (e.g., people, buildings, facilities, equipment, activities, operations, and information) or intangible (e.g., processes or a company's information and reputation).

Asset protection—Security program designed to protect personnel, facilities, and equipment, in all locations and situations, accomplished through planned and integrated application of combating terrorism, physical security, operations security, and personal protective services, and supported by intelligence, counterintelligence, and other security programs.

Asset value—The degree of debilitating impact that would be caused by the incapacity or destruction of an asset.

Audible alarm device—An alarm device that produces an audible announcement (e.g., bell, horn, siren, etc.) of an alarm condition.

Balanced magnetic switch—A door position switch utilizing a reed switch held in a balanced or center position by interacting magnetic fields when not in alarm condition.

Ballistics attack—An attack in which small arms (e.g., pistols, submachine guns, shotguns, and rifles) are fired from a distance and rely on the flight of the projectile to damage the target.

Barbed tape or concertina—A coiled tape or coil of wires with wire barbs or blades deployed as an obstacle to human trespass or entry into an area.

Barbed wire—A double strand of wire with four-point barbs equally spaced along the wire deployed as an obstacle to human trespass or entry into an area.

Barcode—A black bar printed on white paper or tape that can be easily read with an optical scanner.

Biological agents—Living organisms or the materials derived from them that cause disease in or harm to humans, animals, or plants or cause deterioration of material. Biological agents may be used as liquid droplets, aerosols, or dry powders.

Biometric reader—A device that gathers and analyzes biometric features.

Biometrics—The use of physical characteristics of the human body as a unique identification method.

Blast curtains—Heavy curtains made of blast-resistant materials that could protect the occupants of a room from flying debris.

Blast-resistant glazing—Window opening glazing that is resistant to blast effects because of the interrelated function of the frame and glazing material properties frequently dependent upon tempered glass, polycarbonate, or laminated glazing.

Bollard—A vehicle barrier usually consisting of a cylinder, made of steel and sometimes filled with concrete, placed on the ground and spaced about 3 feet apart to prevent vehicles from passing, but allowing entrance of pedestrians and bicycles.

Building hardening—Enhanced construction that reduces vulnerability to external blast and ballistic attacks.

Building separation—The distance between closest points on the exterior walls of adjacent buildings or structures.

Cable barrier—Cable or wire rope anchored to and suspended off the ground or attached to chain-link fence to act as a barrier to moving vehicles.

Capacitance sensor—A device that detects an intruder approaching or touching a metal object by sensing a change in capacitance between the object and the ground.

Card reader—A device that gathers or reads information when a card is presented as an identification method.

Cladding—That portion of the building envelope that provides the weather enclosure. The "skin" of a building.

Clear zone—An area that is clear of visual obstructions and landscape materials that could conceal a threat or perpetrator.

Closed Circuit Television (CCTV)—An electronic system of cameras, control equipment, recorders, and related apparatus used for surveillance or alarm assessment.

CCTV pan-tilt-zoom control—The method of controlling the PTZ functions of a camera.

CCTV pan-tilt-zoom controller—The operator interface for performing PTZ control.

CCTV switcher—A piece of equipment capable of presenting multiple video images to various monitors, recorders, etc.

Confidentiality—The protection of sensitive information against unauthorized disclosure and sensitive facilities from physical, technical, or electronic penetration or exploitation.

Contamination—The undesirable deposition of a chemical, biological, or radiological material on the surface of structures, areas, objects, or people.

Control center—A centrally located room or facility staffed by personnel charged with the oversight of specific situations and/or equipment.

Controlled lighting—Illumination of specific areas or sections.

Controlled perimeter—A physical boundary at which vehicle and personnel access is controlled at the perimeter of a site. Access control at a controlled perimeter should demonstrate the capability to search individuals and vehicles.

Conventional construction—Building construction that is not specifically designed to resist weapons, explosives, or chemical, biological, and radiological effects. Conventional construction is designed only to resist common loadings and environmental effects such as wind, seismic, and snow loads.

Counterintelligence—Information gathered and activities conducted to protect against: espionage, other intelligence activities, sabotage, or assassinations conducted for or on behalf of foreign powers, organizations, or persons; or international terrorist activities, excluding personnel, physical, document, and communications security programs.

Covert entry—Attempts to enter a facility by using false credentials or stealth.

Crash bar—A mechanical egress device located on the interior side of a door that unlocks the door when pressure is applied in the direction of egress. Also called Panic Hardware.

Crime Prevention Through Environmental Design (CPTED)—A crime prevention strategy based on evidence that the design and form of the built environment can influence human behavior. CPTED usually involves the use of three principles: natural surveillance (by placing physical features, activities, and people to maximize visibility); natural access control (through the judicial placement of entrances, exits, fencing, landscaping, and lighting); and territorial reinforcement (using buildings, fences, pavement, signs, and landscaping to express ownership).

Critical assets—Those assets essential to the minimum operations of the organization, and to ensure the health and safety of the general public.

Damage assessment—The process used to appraise or determine the number of injuries and deaths, damage to public and private property, and the status of key facilities and services (e.g., hospitals and other health care facilities, fire and police stations, communications networks, water and sanitation systems, utilities, and transportation networks) resulting from a man-made or natural disaster.

Data transmission equipment—A path for transmitting data between two or more components (e.g., a sensor and alarm reporting system, a card reader and controller, a CCTV camera and monitor, or a transmitter and receiver).

Decontamination—The reduction or removal of a chemical, biological, or radiological material from the surface of a structure, area, object, or person.

Door position switch—A switch that changes state based on whether or not a door is closed—Typically, a switch mounted in a frame that is actuated by a magnet in a door.

Door strike, electronic—An electromechanical lock that releases a door plunger to unlock the door—Typically, an electronic door strike is mounted in place of or near a normal door strike plate. Also called an electric strike.

Dual technology sensor—A sensor that combines two different technologies in one unit.

Duress alarm devices—Also known as panic buttons, these devices are designated specifically to initiate a panic alarm.

Electronic emanations—Electromagnetic emissions from computers, communications, electronics, wiring, and related equipment.

Electronic-emanations eavesdropping—Use of electronic emanation surveillance equipment from outside a facility or its restricted area to monitor electronic emanations from computers, communications, and related equipment.

Electronic Entry Control Systems (EECS)—Electronic devices that automatically verify authorization for a person to enter or exit a controlled area.

Electronic Security System (ESS)—An integrated system that encompasses interior and exterior sensors, closed circuit television systems for assessment of alarm conditions, Electronic Entry Control Systems, data transmission media, and alarm reporting systems for monitoring, control, and display of various alarm and system information.

Emergency—Any natural or human caused situation that results in or may result in substantial injury or harm to the population or substantial damage to or loss of property.

Emergency Medical Services (EMS)—Services including personnel, facilities, and equipment required to ensure proper medical care for the sick and injured from the time of injury to the time of final disposition, including medical disposition within a hospital, temporary medical facility, or special care facility; release from the site; or declared dead. Further, Emergency Medical Services specifically include those services immediately required to ensure proper medical care and specialized treatment for patients in a hospital and coordination of related hospital services.

Emergency Response Plan—A document that describes how people and property will be trained to respond in an emergency or emergency threat situations; details who is responsible for carrying out specific actions; identifies the personnel, equipment, facilities, supplies, and other resources available for use in the emergency; and outlines how actions will be coordinated.

Entry control point—A continuously or intermittently manned station at which entry to sensitive or restricted areas is controlled.

Entry control stations—Entry control stations should be provided at main perimeter entrances where security personnel are present. Entry control stations should be located as close as practical to the perimeter entrance to permit personnel inside the station to maintain constant surveillance over the entrance and its approaches.

Evacuation, organized—Phased, and supervised dispersal of people from dangerous or potentially dangerous areas.

Evacuees—All persons removed or moving from areas threatened or struck by a disaster.

Facial recognition—A biometric technology that is based on features of the human face.

Fence sensor—An exterior intrusion detection sensor that detects aggressors as they attempt to climb over, cut through, or otherwise disturb a fence.

Fiber optics—A method of data transfer by passing bursts of light through a strand of glass or clear plastic.

Field of view—The visible area in a video picture.

Flood—A general and temporary condition of partial or complete inundation of normally dry land areas from overflow of inland or tidal waters, unusual or rapid accumulation or runoff of surface waters, or mudslides/mudflows caused by accumulation of water.

Forced entry—Entry to a denied area achieved through force to create an opening in fence, walls, doors, etc., or to overpower guards.

Fragment retention film (FRF). A thin, optically clear film applied to glass to minimize the spread of glass fragments when the glass is shattered.

Frame rate—In digital video, a measurement of the rate of change in a series of pictures, often measured in frames per second (fps) also sometimes known as the Image Rate or images per second (ips).

Glare security lighting—Illumination projected from a secure perimeter into the surrounding area, making it possible to see potential intruders at a considerable distance while making it difficult to observe activities within the secure perimeter.

Glass-break detector—An intrusion detection sensor that is designed to detect breaking glass either through vibration or acoustics.

Glazing—A material installed in a sash, ventilator, or panes (e.g., glass, plastic, etc., including material such as thin granite installed in a curtain wall).

Grid wire sensor—An intrusion detection sensor that uses a grid of wires to cover a wall or fence. An alarm is sounded if the wires are cut.

Hand geometry—A biometric technology that is based on characteristics of the human hand.

Hazard—A source of potential danger or adverse condition.

Hazard mitigation—Any action taken to reduce or eliminate the long-term risk to human life and property from hazards. The term is sometimes used in a stricter sense to mean cost-effective measures to reduce the potential for damage to a facility or facilities from a disaster event.

Hazardous material (HazMat)—Any substance or material that, when involved in an accident and released in sufficient quantities, poses a risk to people's health, safety, and/or property. These substances and materials include explosives, radioactive materials, flammable liquids or solids, combustible liquids or solids, poisons, oxidizers, toxins, and corrosive materials.

Human-caused hazard—Human-caused hazards are technological hazards and terrorism. They are distinct from natural hazards primarily in that they originate from human activity. Within the military services, the term threat is typically used for human-caused hazard.

Hurricane—A tropical cyclone, formed in the atmosphere over warm ocean areas, in which wind speeds reach 74 miles per hour or more and blow in a large spiral around a relatively calm center or "eye." Circulation is counter-clockwise in the Northern Hemisphere and clockwise in the Southern Hemisphere.

Impact analysis—A management level analysis that identifies the impacts of losing the entity's resources. The analysis measures the effect of resource loss and escalating losses over time in order to provide the entity with reliable data upon which to base decisions on hazard mitigation and continuity planning.

Incident Command System (ICS)—A standardized organizational structure used to command, control, and coordinate the use of resources and personnel that have responded to the scene of an emergency. The concepts and principles for ICS include common terminology, modular organization, integrated communication, unified command structure, consolidated action plan, manageable span of control, designated incident facilities, and comprehensive resource management.

Insider compromise—A person authorized access to a facility (an insider) compromises assets by taking advantage of that accessibility.

Intercom door/gate station—Part of an intercom system where communication is typically initiated, usually located at a door or gate.

Intercom master station—Part of an intercom system that monitors one or more intercom door/gate stations; typically, where initial communication is received.

Intercom switcher—Part of an intercom system that controls the flow of communications between various stations.

Intercom system—An electronic system that allows simplex (one way), half-duplex (one way and answer one way), or full-duplex (both talk and listen at the same time) audio communications.

Intrusion Detection Sensor—A device that initiates alarm signals by sensing the stimulus, change, or condition for which it was designed.

Intrusion Detection System (IDS)—The combination of components, including sensors, control units, transmission lines, and monitor units, integrated to detect intrusion and respond in a specified manner.

Isolated fenced perimeters—Fenced perimeters with 100 feet or more of space outside the fence that is clear of obstruction, making approach obvious.

Jersey barrier—A protective concrete barrier used as a highway divider that now also functions as an expedient method for traffic speed control at entrance gates and to keep vehicles away from buildings.

Joint Information Center (JIC)—A central point of contact for all news media near the scene of a large scale disaster. News media representatives are kept informed of activities and events by Public Information Officers who represent all participating federal, state, and local agencies that are collocated at the JIC.

Joint Operations Center (JOC)—The focal point for management and direction of on-site activities, coordination/establishment of state requirements/priorities, and coordination of the overall response.

Jurisdiction—Typically counties and cities within a state, but states may elect to define differently in order to facilitate their assessment process.

Laminated glass—A plate of glass of uniform thickness consisting of two monolithic glass plies bonded together with an interlayer material as defined in ASTM Specification C-1172. Many different interlayer materials are used in laminated glass.

Landscaping—The use of plantings (shrubs and trees), with or without landforms and/or large boulders, to act as a perimeter barrier against defined threats.

Laser card—A card technology that uses a laser reflected off of a card for uniquely identifying the card.

Layers of protection—A traditional approach in security engineering using concentric circles extending out from an area to be protected as demarcation points for different security strategies.

Line of sight (LOS)—Direct observation between two points with the naked eye or hand-held optics.

Line-of-sight sensor—A pair of devices used as an intrusion detection sensor that monitor any movement through the field between the sensors.

Line supervision—A data integrity strategy that monitors the communications link for connectivity and tampering. In Intrusion Detection System sensors, line supervision is often referred to as two-state, three-state, or four-state in respect to the number of conditions monitored. The frequency of sampling the link also plays a big part in the supervision of the line.

Local government—Any county, city, village, town, district, or political subdivision of any state, and Indian tribe or authorized tribal organization, or Alaska Native village or organization, including any rural community or unincorporated town or village or any other public entity.

Magnetic lock—An electromagnetic lock that unlocks a door when power is removed.

Magnetic stripe—A card technology that uses a magnetic stripe on the card to encode data used for unique identification of the card.

Man-trap—An access control strategy that uses a pair of interlocking doors to prevent tailgating. Only one door can be unlocked at a time.

Microwave motion sensor—An intrusion detection sensor that uses microwave energy to sense movement within the sensor's field of view. These sensors work similar to radar by using the Doppler effect to measure a shift in frequency.

Minimum measures—Protective measures that can be applied to all buildings regardless of the identified threat. These measures offer defense or detection opportunities for minimal cost, facilitate future upgrades, and may deter acts of aggression.

Mitigation—Those actions taken to reduce the exposure to and impact of an attack or disaster.

Motion detector—An intrusion detection sensor that changes state based on movement in the sensor's field of view.

Moving vehicle bomb—An explosive-laden car or truck for the purpose of being driven into or near a building and detonated.

Natural hazard—Naturally-occurring events such as floods, earthquakes, tornadoes, tsunami waves, coastal storms, landslides, and wildfires that strike populated areas. A natural event is a hazard when it has the potential to harm people or property. The risks of natural hazards may be increased or decreased as a result of human activity; however, they are not inherently human-induced.

Natural protective barriers—Natural protective barriers are mountains and deserts, cliffs and ditches, water obstacles, or other terrain features that are difficult to traverse.

Nuclear, biological, or chemical weapons. Also called Weapons of Mass Destruction (WMD). Weapons that are characterized by their capability to produce mass casualties.

On-scene Incident Commander—The official designated upon event activation to ensure appropriate coordination of the overall response. This individual has authority to make life/death decisions for responders.

Open systems architecture—A term borrowed from the information technology industry to claim that systems are capable of interfacing with other systems from any vendor, which also uses open system architecture. The opposite would be a proprietary system.

Operator interface—The part of a security management system that provides user interface to humans.

Organizational areas of control—Controls consist of the policies, procedures, practices, and organization structures designed to provide reasonable assurance that business objectives will be achieved and that undesired events will be prevented or detected and corrected.

Pan-tilt-zoom camera (PTZ)—A CCTV camera that can move side to side, up and down, and zoom in or out.

Passive infrared motion sensor (PIR)—A device that detects a change in the thermal energy pattern caused by a moving intruder and initiates an alarm when the change in energy satisfies the detector's alarm-criteria.

Passive vehicle barrier—A vehicle barrier that is permanently deployed and does not require response to be effective.

Perimeter barrier—A fence, wall, vehicle barrier, land form, or line of vegetation applied along an exterior perimeter used to obscure vision, hinder personnel access, or hinder or prevent vehicle access.

Physical security—The part of security concerned with measures/concepts designed to safeguard personnel; to prevent unauthorized access to equipment, installations, materiel, and documents; and to safeguard them against espionage, sabotage, damage, and theft.

Planter barrier—A passive vehicle barrier, usually constructed of concrete and filled with dirt (and flowers for aesthetics). Planters, along with bollards, are the usual street furniture used to keep vehicles away from existing buildings. Overall size and the depth of installation below grade determine the vehicle stopping capability of the individual planter.

Polycarbonate glazing—A plastic glazing material with enhanced resistance to ballistics or blast effects.

Preparedness—Establishing the plans, training, exercises, and resources necessary to enhance mitigation of and achieve readiness for response to, and recovery from all hazards, disasters, and emergencies, including WMD incidents.

Pressure mat—A mat that generates an alarm when pressure is applied to any part of the mat's surface, such as when someone steps on the mat. Pressure mats can be used to detect an intruder approaching a protected object, or they can be placed by doors and windows to detect entry.

Primary asset—An asset that is the ultimate target for compromise by an aggressor.

Probability of detection (POD)—A measure of an intrusion detection sensor's performance in detecting an intruder within its detection zone.

Probability of intercept—The probability that an act of aggression will be detected and that a response force will intercept the aggressor before the asset can be compromised.

Progressive collapse—A chain reaction failure of building members to an extent disproportionate to the original localized damage. Such damage may result in upper floors of a building collapsing onto lower floors.

Protective barriers—Define the physical limits of a site, activity, or area by restricting, channeling, or impeding access and forming a continuous obstacle around the object.

Protective measures—Elements of a protective system that protect an asset against a threat. Protective measures are divided into defensive and detection measures.

Protective system—An integration of all of the protective measures required to protect an asset against the range of threats applicable to the asset.

Proximity sensor—An intrusion detection sensor that changes state based on the close distance or contact of a human to the sensor. These sensors often measure the change in capacitance as a device enters the measured field.

Public Information Officer (PIO)—A federal, state, or local government official responsible for preparing and coordinating the dissemination of emergency public information.

Recovery—The long-term activities beyond the initial crisis period and emergency response phase of disaster operations that focus on returning all systems in the community to a normal status or to reconstitute these systems to a new condition that is less vulnerable.

Request-to-exit device—Passive infrared motion sensors or push buttons that are used to signal an Electronic Entry Control System that egress is imminent or to unlock a door.

Resolution—The level to which video details can be determined in a CCTV scene. Also called the resolving ability.

Response—Executing the plan and resources identified to perform those duties and services to preserve and protect life and property as well as provide services to the surviving population.

Response force—The people who respond to an act of aggression. Depending on the nature of the threat, the response force could consist of guards, special reaction teams, military or civilian police, an explosives ordnance disposal team, or a fire department.

Response time—The length of time from the instant an attack is detected to the instant a security force arrives on site.

Retinal pattern—A biometric technology that is based on features of the human eye.

RF data transmission—A communications link using radio frequency to send or receive data.

Risk—The potential for loss of, or damage to, an asset. It is measured based upon the value of the asset in relation to the threats and vulnerabilities associated with it.

Routinely occupied—For the purposes of these standards, an established or predictable pattern of activity within a building that terrorists could recognize and exploit.

RS-232 data—IEEE Recommended Standard 232; a point-to-point serial data protocol with a maximum effective distance of 50 feet.

RS-422 data—IEEE Recommended Standard 422; a point-to-point serial data protocol with a maximum effective distance of 4,000 feet.

RS-485 data—IEEE Recommended Standard 485; a multi-drop serial data protocol with a maximum effective distance of 4,000 feet.

Safe haven—Secure areas within the interior of the facility. A safe haven should be designed such that it requires more time to penetrate by aggressors than it takes for the response force to reach the protected area to rescue the occupants. It may be a haven from a physical attack or air-isolated haven from CBR contamination.

Secondary asset—An asset that supports a primary asset and whose compromise would indirectly affect the operation of the primary asset.

Secondary hazard—A threat whose potential would be realized as the result of a triggering event that of itself would constitute an emergency (e.g., dam failure might be a secondary hazard associated with earthquakes).

Secure/access mode—The state of an area monitored by an intrusion detection system in regards to how alarm conditions are reported.

Security analysis—The method of studying the nature of and the relationship between assets, threats, and vulnerabilities.

Security console—Specialized furniture, racking, and related apparatus used to house the security equipment required in a control center.

Security engineering—The process of identifying practical, risk managed short and long term solutions to reduce and/or mitigate dynamic man-made hazards by integrating multiple factors, including construction, equipment, manpower, and procedures.

Security engineering design process—The process through which assets requiring protection are identified, the threat to and vulnerability of those assets is determined, and a protective system is designed to protect the assets.

Semi-isolated fenced perimeters—Fence lines where approach areas are clear of obstruction for 60 to 100 feet outside of the fence and where the general public or other personnel seldom have reason to be in the area.

Serial interface—An integration strategy for data transfer where components are connected in series.

Shielded wire—Wire with a conductive wrap used to mitigate electromagnetic emanations.

Smart card—A newer card technology that allows data to be written, stored, and read on a card typically used for identification and/or access.

Specific threat—Known or postulated aggressor activity focused on targeting a particular asset.

Stand-off distance—A distance maintained between a building or portion thereof and the potential location for an explosive detonation or other threat.

Stationary vehicle bomb—An explosive-laden car or truck

stopped or parked near a building.

Storm surge—A dome of sea water created by the strong winds and low barometric pressure in a hurricane that causes severe coastal flooding as the hurricane strikes land.

Strain sensitive cable—Strain sensitive cables are transducers that are uniformly sensitive along their entire length and generate an analog voltage when subjected to mechanical distortions or stress resulting from fence motion. They are typically attached to a chain-link fence about halfway between the bottom and top of the fence fabric with plastic ties.

Structural protective barriers—Man-made devices (e.g., fences, walls, floors, roofs, grills, bars, roadblocks, signs, or other construction) used to restrict, channel, or impede access.

Superstructure—The supporting elements of a building above the foundation.

System events—Events that occur normally in the operation of a security management system. Examples include access control operations and changes of state in intrusion detection sensors.

Tamper switch—Intrusion detection sensor that monitors an equipment enclosure for breach.

Tangle-foot wire—Barbed wire or tape suspended on short metal or wooden pickets outside a perimeter fence to create an obstacle to approach.

Taut wire sensor—An intrusion detection sensor utilizing a column of uniformly spaced horizontal wires, securely anchored at each end and stretched taut. Each wire is attached to a sensor to indicate movement of the wire.

TEMPEST—An unclassified short name referring to investigations and studies of compromising emanations. It is sometimes used synonymously for the term "compromising emanations" (e.g., TEMPEST tests, TEMPEST inspections).

Terrorism—The unlawful use of force and violence against persons or property to intimidate or coerce a government, the civilian population, or any segment thereof, in furtherance of political or social objectives.

Thermally tempered glass (TTG)—Glass that is heat-treated to have a higher tensile strength and resistance to blast pressures, although with a greater susceptibility to airborne debris.

Threat—Any indication, circumstance, or event with the potential to cause loss of, or damage to, an asset.

Threat analysis—A continual process of compiling and examining all available information concerning potential threats and human-caused hazards. A common method to evaluate terrorist groups is to review the factors of existence, capability, intentions, history, and targeting.

Time/date stamp—Data inserted into a CCTV video signal with the time and date of the video as it was created.

TNT equivalent weight—The weight of TNT (trinitrotoluene) that has an equivalent energetic output to that of a different weight or another explosive compound.

Tornado—A local atmospheric storm, generally of short duration, formed by winds rotating at very high speeds, usually in a counter-clockwise direction. The vortex, up to several hundred yards wide, is visible to the observer as a whirlpool-like column of winds rotating about a hollow cavity or funnel. Winds may reach 300 miles per hour or higher.

Toxicity—A measure of the harmful effects produced by a given amount of a toxin on a living organism.

Triple-standard concertina (TSC) wire—This type of fence uses three rolls of stacked concertina—One roll will be stacked on top of two other rolls that run parallel to each other while resting on the ground, forming a pyramid.

Tsunami—Sea waves produced by an undersea earthquake. Such sea waves can reach a height of 80 feet and can devastate coastal cities and low-lying coastal areas.

Two-person rule—A security strategy that requires two people to be present in, or gain access to, a secured area to prevent unobserved access by any individual.

Unobstructed space—Clear space around an inhabited building without an obstruction large enough to conceal an explosive device, greater than 150 mm (6 inches) in height.

Unshielded wire—Wire that does not have a conductive wrap.

Unshielded Twisted Pair Wire—Wire that uses pairs of wires twisted together to mitigate electromagnetic interference.

Vault—A reinforced room for securing items.

Vertical rod—Typical door hardware often used with a crash bar to lock a door by inserting rods vertically from the door into the doorframe.

Vibration sensor—An intrusion detection sensor that changes state when vibration is present.

Video intercom system—An intercom system that also incorporates a small CCTV system for verification.

Video motion detection—Motion detection technology that looks for changes in the pixels of a video image.

Video multiplexer—A device used to connect multiple video signals to a single location for viewing and/or recording.

Visual displays—A display or monitor used to inform the operator visually of the status of the electronic security system.

Visual surveillance—The aggressor uses ocular and photographic devices (such as binoculars and cameras with telephoto lenses) to monitor facility or installation operations or to see assets.

Voice recognition—A biometric technology that is based on nuances of the human voice.

Volumetric motion sensor—An interior intrusion detection sensor that is designed to sense aggressor motion within a protected space.

Vulnerability—Any weakness that can be exploited by an aggressor or, in a non-terrorist threat environment, make an asset susceptible to hazard damage.

Warning—The alerting of emergency response personnel and the public to the threat of extraordinary danger and the related effects that specific hazards may cause.

Watch—Indication in a defined area that conditions are favorable for the specified type of severe weather (e.g., flash flood watch, severe thunderstorm watch, tornado watch, tropical storm watch).

Waterborne contamination—Chemical, biological, or radiological agent introduced into and fouling a water supply.

Weigand protocol—A security industry standard data protocol for card readers.

Zoom—The ability of a CCTV camera to close and focus or open and widen the field of view.

Index

325